Health Travels

Cuban Health(care) On and Off the Island

Perspectives in Medical Humanities

Perspectives in Medical Humanities publishes scholarship produced or reviewed under the auspices of the University of California Medical Humanities Consortium, a multi-campus collaborative of faculty, students and trainees in the humanities, medicine, and health sciences. Our series invites scholars from the humanities and health care professions to share narratives and analysis on health, healing, and the contexts of our beliefs and practices that impact biomedical inquiry.

General Editor

Brian Dolan, PhD, Professor of Social Medicine and Medical Humanities, University of California, San Francisco (UCSF)

Recent Titles

Clowns and Jokers Can Heal Us: Comedy and Medicine
By Albert Howard Carter III (Fall 2011)

The Remarkables: Endocrine Abnormalities in Art
By Carol Clark and Orlo Clark (Winter 2011)

Health Citizenship: Essays in Social Medicine and Biomedical Politics
By Dorothy Porter (Winter 2011)

*What to Read on Love, not Sex: Freud, Fiction, and the Articulation
of Truth in Modern Psychological Science*
By Edison Miyawaki, MD; Foreword by Harold Bloom (Fall 2012)

Patient Poets: Illness from Inside Out
By Marilyn Chandler McEntyre (Spring 2013)

www.UCMedicalHumanitiesPress.com

brian.dolan@ucsf.edu

This series is made possible by the generous support of the Dean of the School of Medicine at UCSF, the Center for Humanities and Health Sciences at UCSF, and a MCRP grant from the University of California Office of the President.

Health Travels

Cuban Health(care) On and Off the Island

Edited by Nancy J. Burke, PhD

First published in 2013
University of California Medical Humanities Press in partnership with
California Digital Library
San Francisco - Berkeley

© 2013
University of California
Medical Humanities Consortium
3333 California Street, Suite 485
San Francisco, CA 94143-0850

Designed by Eduardo de Ugarte

Library of Congress Control Number: 2013937068

ISBN 978-0-9889865-1-0

Printed in USA

Contents

Acknowledgments

This volume emerged from an October 2010 conference I organized with Raul Fernandez, Director of the University of California (UC)-Cuba Multi-Campus Academic Initiative, in Berkeley, CA entitled "In Sickness and in Health: Encountering Wellness in Cuba and the United States." The many hours we spent planning the conference, and subsequently discussing this volume, were unimaginable when we first came up with the idea in the café of the *Habana Libre* over *cortados*. I am indebted to Raul for his mentorship, humor, and calm presence. Special thanks go to Laura Enriquez, Lisa García-Bedolla, Kamran Nayeri, and Richard Quint for their help planning and organizing the conference. Funding to bring together U.S.- and Cuba-based scholars in the conference was provided by the UC-Cuba Multi-Campus Academic Initiative, the UCSF Global Health Sciences Health Diplomacy and Medical Education in Cuba Program, the UCSF/UCB Joint Medical Anthropology Program, the UC Institute on Global Conflict and Cooperation, the UCB Townsend Center for the Humanities, the UCB Center for Latino Policy Research, and the UCB Andrew W. Mellon Program in Latin American Sociology. A subset of the papers presented are included herein, and have been discussed and presented in annual meetings of the American Anthropological Association. Thanks to all the contributors for their collegiality, insight, originality, and commitment to critical engagement. Appreciation is owed to three anonymous reviewers who carefully read each chapter and offered wise and constructive suggestions for revision. Hannah Leslie assisted in the preparation of the manuscript. Without her help and careful eye for detail I would possibly still be checking bibliographies. Thanks also go to the series editor, Brian Dolan, for his support and guidance through this process. Special appreciation for Tony, Marina, Nibaldo, Juan Carlos, Caceres, Elba, and Javier who have taught me so much about health and care on the island, and to Niurka and Xiomara who have welcomed me into their homes and lives over the last twenty years. Finally, thanks go to the faculty and graduate students of the UC-Cuba Multi-Campus Academic Initiative whose excitement, interest, and questions in annual workshops and meetings encouraged me to continue this work.

Nancy J. Burke

Contributors

Elise Andaya is Assistant Professor of Anthropology at the University at Albany (SUNY). She has published in journals such as *Medical Anthropology Quarterly*, *Feminist Studies*, as well as in a number of edited volumes, and is currently finishing a book that examines shifts in ideologies and practices around reproduction, gender, and kinship in Havana since the fall of the socialist bloc.

Alissa Bernstein is a PhD candidate in the Joint University of California, San Francisco/University of California, Berkeley Medical Anthropology Program. Her dissertation research is on the designing and circulation of new health reform policy in Bolivia. She has also conducted research in Cuba and Bolivia on medical education and medical humanitarianism.

Charles L. Briggs is a Professor in the Department of Anthropology, University of California, Berkeley and the Joint University of California, San Francisco/University of California, Berkeley Medical Anthropology Program. He has published widely in anthropology and social science journals. His books include *Learning How to Ask* (1986), *Voices of Modernity* (with Richard Bauman, 2003), and *Stories in the Time of Cholera* (with Clara Mantini-Briggs, 2004).

P. Sean Brotherton is an Assistant Professor of Anthropology at Yale University. His research and teaching interests are concerned with the critical study of health, medicine, the state, subjectivity, and the body, drawing upon contemporary social theory and postcolonial studies. He conducts fieldwork in the Caribbean and Latin American, particularly Cuba and Jamaica, and is the author of *Revolutionary Medicine: Health and the Body in Post-Soviet Cuba* (2013).

Nancy J. Burke is an Associate Professor in the Department of Anthropology, History, and Social Medicine at the University of California, San Francisco and the Joint University of California, San Francisco/University of California, Berkeley Medical Anthropology Program. Since 2009 she has served as the Director of the Health Diplomacy and Medical Education in Cuba Program in the UCSF Institute of Global Health Sciences. Her research focuses on the anthropology of cancer in the United States and chronic disease management in Cuba.

Susannah Rodríguez Drissi is a lecturer in the Department of Comparative Literature, University of California, Los Angeles. Her dissertation research proposes an interpretive paradigm to explore the Moorish, Arab, Islamic, and Algerian presence in Cuban literature and cultural studies. She is currently completing an English translation of Cuban ethnographer Lydia Cabrera's *Porque...Cuentos negros de Cuba* [Because...Afro-Cuban Tales] with critical introduction, to be published by University Press of Florida.

Julie Feinsilver is a Scholar-in-Residence at the American University School of International Service's International Development Program, and an independent consultant. She has conducted research on Cuban medical diplomacy since 1979. Her publications include *Healing the Masses: Cuban Health Politics At Home and Abroad* (1993) as well as numerous articles and book chapters addressing Cuba's medical diplomacy, biotechnology development, foreign relations, non-traditional exports, and politics of health.

Hanna Garth is a PhD candidate in the Department of Anthropology, University of California, Los Angeles. She is broadly interested in studying how people use food systems. Her most recent project analyzes the ways in which the changing food rationing system in Cuba affects household and community dynamics and in turn individual subject positions. She also conducts fieldwork in Los Angeles-based food justice organizations, and has previously worked in the Philippines, Chile, Peru and Houston, TX. Ms. Garth is the editor of the volume *Food and Identity in the Caribbean* (2013).

Sahra Gibbon is a Lecturer in the Anthropology Department at University College London. She has carried out research in the United Kingdom, Cuba and Brazil and is the author of a number of monographs and articles examining the social and cultural context of developments in genetic knowledge and technology, including *Breast Cancer Genes and the Gendering of Knowledge* (2006) and *Biosocialities, Genetics and the Social Sciences; Making Biologies and Identities* (with Carlos Novas, 2008).

Ayesha Anne Nibbe is an Assistant Professor of Anthropology at Hawaii Pacific University. She conducts research on Development and Humanitarianism in Africa, and is currently producing a book manuscript on the international humanitarian response in northern Uganda. While working on her doctoral research, she became acquainted with a group of Cuban *Internacionalistas* working in Uganda who ultimately expanded her research on international aid in Africa.

Julio César González Pagés is a Professor in the Department of History and Philosophy, University of Havana. He is General Coordinator of *Red Iberoamericana de Masculinidades* (Iberoamerican Masculinities Network) and works as consultant on masculinity and violence in Latin America for various organizational branches of the United Nations. He is an active member of the Academy of Cuban History and has published numerous articles on the history of the feminist movement in Cuba. His most recent book is *Macho, varón, masculino. Estudios de masculinidades en Cuba* (2010).

Introduction: Cuban Health(care) and the Anthropology of Global Health

Nancy J. Burke

The contentiousness and complexity of the subject of this book – the Cuban health care system – became clear to me in the early evening of a sweltering June day in 1997 as I swerved around potholes in a Soviet-model Lada, heading toward a large urban hospital in the Vedado neighborhood of Havana. My friend and her 18-month-old son were in the back seat; he whimpering, she looking concerned. The little boy had had severe asthma since his birth. Their morning ritual since his first attack included waiting in line outside the neighborhood *consultorio* (primary care clinic) for his turn with the nebulizer. That morning he had woken up with hives all over his body. The family doctor recommended that he be taken to the hospital as the reaction seemed to intensify over the course of the day. The previous three hours had been spent scrambling; collecting sheets, buying cotton from the dollar store, finding alcohol and the list of other supplies families were required to bring for a hospital stay.

This was my second summer spent in Havana in the midst of what then-President Fidel Castro had deemed the "Special Period in a Time of Peace."[1] The moniker was used to describe the institution of wartime rations in response to the collapse of the Soviet Union in 1989. Having spent three months on the island in 1994, the summer over 21,000 *balseros* (rafters) were intercepted at sea trying to make it to the United States, I was familiar with the constraints of the "Special Period"; the brown-outs, black-outs, limitations of *la libreta* (ration book), the soft knocks at odd hours signaling the arrival of black-marketeers offering eggs or oil no longer available in the state-run stores, the image of entire families riding on a single bicycle due to the lack of gasoline, and the virtual absence of vegetables in urban centers. I had met people who were raising pigs in their bathtubs, students who had dropped out of school to pursue hard currency through relationships with tourists, and young men who gained new power through their unofficial positions as intermediaries with foreign businesses interested in establishing *inversiones* (joint ventures) on the island. I had not previously, however, entered a Cuban hospital.

My exposure to Cuban health care up to that point had included hearing from women walking around with acupuncture needles protruding from their upper ears that this helped address chronic headaches brought on by the stress of *inventando* (trying to make do with less), and participation in a number of *limpias* (cleansings) and consultations performed by *santeros* (priests) in the Regla de Ocha tradition,[2] at times to ascertain the veracity of claims of HIV infection. Many of the people I knew made comments reflecting their ambivalence with the state of the health care system at the time, such as "one of the triumphs of the revolution is that our health care is free. I can see a doctor any time I want. But once I get there, he has nothing to give me," referring to the lack of pharmaceuticals available after the collapse of the Soviet Union and the strengthening of the U.S. trade embargo with the passage of the Torricelli Amendment in 1992 and the Helms-Burton Act in 1996.[3] Such comments took on new meaning when I entered the hospital that day and worked with my friend to create a warm and safe seeming space for her son in the midst of dirty floors and walls suffering from disrepair as paint pealed from their surfaces.

The contradictions inherent in this experience – the value of free universal access to care and the challenging context of that care – resound throughout the chapters included in this volume. These contributions stem from a 2010 conference at the University of California, Berkeley, entitled "In Sickness and Health: Encountering Wellness in Cuba and the United States." The one-day conference brought together U.S. and Cuban scholars from medical anthropology, sociology, political science, and epidemiology to engage in critical dialogue about the Cuban health care system, often held up as a symbol of revolutionary success and counter-hegemonic possibility for health practitioners, scholars, journalists, and politicians throughout the world. Participants in the conference questioned static representations of the Cuban health care system and argued for ethnographically grounded analysis of the many forms this system has taken on and off the island. Rather than taking as a given the value and exceptionalism of Cuban health care, therefore, contributors to this volume ethnographically explore how this system, and its many successes, are produced, as well as the (un)intended consequences of its production and maintenance felt and experienced on the island and around the world.

The purpose of this introduction is threefold: to provide a brief window into social and economic transformations with implications for health – such as those referenced above – during and since the "Special Period," to situate study of the Cuban health care system within current debates in the anthropology of global health, and to briefly introduce the chapters. The thrust of the introduction is an argument for the relevance of anthropological study of Cuban health(care) to furthering our understandings of "the local," following Ginsburg and Rapp's definition as being "not defined by geographical boundaries but understood as any small-scale area in which social meanings are informed and adjusted" (1995: 8) within the context of reconfigurations of global relations in the post-Cold War era. Anthropological work in global health requires a focus on the instantiation of global assemblages in local social arenas (Janes and Corbett 2009). The task taken up by contributors to this volume is to understand how various assemblages of global, national, and subnational factors converge on health issues, problems, or outcomes in the particular local context of Cuban health(care) (Janes and Corbett 2009).

Previous work by medical anthropologists and sociologists on the island has generated critical historical analyses of the health care system (Crabb 2001; Hirschfeld 2007), laudatory accounts of socialized medicine (Whiteford and Branch 2009), and in-depth reviews of Cuba's program of medical diplomacy (Feinsilver 1993; Brouwer 2011; J. Kirk and Erisman 2009). Most recently, P. Sean Brotherton's ethnography of Cuban healthcare in the 1990s and early 2000s details how changes in the system have been lived and experienced on the island (Brotherton 2012). *Health Travels* is the first collection of its kind to bring together in-depth critical analyses of elemental aspects of Cuban health(care), spanning from a dengue epidemic to food insecurity, medical diplomacy, reproductive health, men's health, and the emergence of "community genetics."

The Special Period and its Aftereffects

The deprivations briefly outlined in the ethnographic vignette opening this introduction stemmed from the dissolution of the Soviet Union and the Community for Economic Cooperation (COMECON) in 1989, which resulted in the reduction of Soviet Bloc exports to Cuba by about 70% by 1993. In the same time span, the value of all Cuban imports declined from $8 billion to

1.7 billion (Maal-Bared 2006; Uriarte 2005). Prior to the dissolution, trade with countries in the bloc had accounted for over 85% of commerce, and 98% of Cuba's electricity production had been fossil fuel dependent (Eckstein 2010). The collapse of its main trade partners meant the island was left without fossil fuel. Many factories closed and the transportation system was crippled. Agricultural production, modeled after the Green Revolution, was practically stopped due to lack of fertilizers, weed killers, and fuel and parts for irrigation pumps. This resulted in serious food shortages and a decrease of 30% of caloric intake between 1990 and 1995 (Uriarte 2005). Implications for health included disrepair of hospital structures and equipment (De Vos et al. 2010), nutritional deficiencies, reemergence of previously eradicated epidemics, increases in low-birthweight babies, and lack of pharmaceuticals. In response, the government promoted tourism, curtailed consumption, cut back state spending (but maintained proportional investment in social services including health and education), implemented a food self-sufficiency program, legalized foreign investment and possession of U.S. dollars, and invested in green medicine and biotechnology (Maal-Bared 2006; Uriarte 2005).

During the Special Period, food and merchandise costs increased 28% and 50% respectively, production costs increased 30%, and shipping costs increased up to 400% (Fernandez and Angeles 2009). Wages remained the same until the increase in the mid 2000s,[4] and many people faced unemployment for the first time since 1959. In 1997 the percentage of men unemployed reached over four percent and of women over ten percent (Uriarte 2005). The burden of scarcities was gendered as women continued to take care of domestic needs with less resources and state support. Thus, in addition to working outside the home and in the community, women also spent more time waiting in line at state-run stores, searching for difficult to find items, and waiting for unreliable transportation (Fernandez and Angeles 2009). Referred to colloquially as *Plan Jaba*, reductions in consumption meant that no one left the house without at least a plastic bag (*jaba*) just in case they ran into someone selling on the street.

The government's strategy of dollar legalization in 1993[5] resulted in the emergence of tangible income differentials across the island. Particularly for those with family abroad, access to resources and dollars via remittances meant better housing, nutrition, and clothing – the basic needs that had degraded since 1989. Ramifications of this income inequality were wide ranging. In the 1990s debates about the implications of the dual economy and what was often referred to as tourist apartheid (Mazzei 2012) were held on street corners and at academic conferences in Cuba and the United States. Between the summers of 1997 and 1998, the effects on Havana streets were striking: select houses received new coats of paint and fences appeared around them, indicating that they had family abroad, while others fell into disrepair. Similarly, investments in the tourism sector meant some streets in Old Havana were renovated, while others, where small artists studios and *botanicas*[6] had eked out a living catering to adventurous tourists, were left "off the map," and thus lost access to previous sources of hard currency. Cruise ships docking in the port provided tourists with printed guides to keep them on newly renovated tourist routes. Throughout the 1990s and 2000s, religious, cultural, medical, and educational tourism flourished, bringing in much needed hard currency and often unanticipated social transformations (Roland 2010; Cabezas 2009).

Tourism turned professional hierarchies in place since 1959 on their head. Prior to 1989, doctors or engineers were the highest paid workers, earning almost five times as much as the lowest paid workers. During the 1990s, it became possible for a waiter in a tourist hotel to earn many times more than a top professional due to their access to hard currency via tips, and the "extras" available to them through the tourist sector. Referred to by Cuba watchers as the "inverted pyramid," implications for power relations within families – where young men and women working in tourism earned much more than parents working in professional government jobs – were profound (Uriarte 2005) and challenged one of the most oft-touted triumphs of the Revolution: universal access to higher education.

Importantly, the new income inequality resulted in a new dietary inequality (Eckstein 2010). As mentioned at the opening of this Introduction, all Cubans are entitled to state-supplied rations via *la libreta*, but the prices at private farmers' markets, legalized in 1994, and state stores, often the only places where vegetables and ingredients such as garlic and onions, beef and pork – those necessary to give food the *gusto* (flavor) considered elemental to Cuban cuisine – were often ten to 100 times higher than prices regulated through the ration system. Therefore, hard currency accessed via remittances, tourism, and/or black market activity provided access to dietary options not available to the peso dependent. Further, remittances and hard currency accessed through work in tourism exacerbated race-based inequality as those sending remittances were overwhelmingly white, as were those able to secure positions in tourist hotels and restaurants (Eckstein 2010; Alzugaray Treto 2009).[7] The implications of these dietary differences for mental health and *Cubanidad* (Cuban identity) are explored in both Hanna Garth's and Susannah

Drissi's contributions to this volume.

Another indication of health impacts of the deprivations of the Special Period was the increase in reported low-birthweight babies and decreased nutrition among young children (Uriarte 2005). By the time of the collapse of the Soviet trading bloc in 1989, infant mortality was 11 out of 1000, the lowest in Latin America, and despite the challenges of the subsequent years, reached 9.3 in 1993 (Tancer 1995). However, during the same period, the prevalence of babies born with low weight (under 2500 grams) began to increase. The rate in 1993 reached 9% compared to 8.7% in 1988 (Uriarte 2005). In response, the National Program of Low Birth Weights worked with the Ministry of Public Health (MINSAP) and local governments to identify and address the needs of at-risk pregnant women. At the neighborhood level, as detailed in Elise Andaya's contribution to this volume, at-risk women were identified and sent to *hogares maternos* (maternal homes), where they were monitored and provided nutritional supplementation. By 1995, these and other efforts to improve nutritional support for pregnant women (e.g. connecting them with worker's lunchrooms for free meals) resulted in a reduction of low-birthweight births to just under eight percent. Recent reports indicate that Cuba ended 2011 with an infant mortality rate of 4.9 per 1000 live births ('Cuba's Child Mortality' 2012).[8] For the previous four years, Cuba had topped the list of countries on the continent with less than five percent infant mortality. Andaya's chapter details the daily work devoted, and the stress induced by the need, to bring down those numbers.

Social responses to the scarcities experienced in the 1990s and fears for the future included mass migration to the United States. In August 1994, 21,300 Cubans were intercepted at sea on makeshift rafts and inner tubes, with a decrease to approximately 11,000 in September (LAI and CEA 1995). In response, in 1995 the United States' "open-door" immigration policy in effect since 1959 was transformed into a "wet foot, dry foot" policy, or conditional entrance and repatriation of those rescued at sea. Once a Cuban arrived on U.S. shores, she gained unconditional entrance as a parolee. If intercepted at sea, prior to touching "dry" land, she was returned to the island. Cuban migration to the United States was further limited through an immigration lottery, *el bombo*, held by the U.S. Interests Office on the island. The goal of this lottery was to regulate Cuban – U.S. immigration to 20,000 people per year. Migration in the 1990s transformed the Cuban population in the United States (see Gosin 2010; Burke 2001).

The 2000s have seen the gradual end of the "Special Period" and continual rapid economic, social, and political change on the island. Changes range from the transition of state power from Fidel to his brother Raúl Castro to increasing entrepreneurism, including the ability to buy and sell property (Cave 2011). In the health arena, changes have included official discussions and conferences on homosexuality and bisexuality (Reed 2012), the development and offering of the first sex-change surgery (Rodríguez 2010), innovation in biotech development from cancer therapeutics to diabetes treatments (Berlanga et al. 2013), decentralization and restructuring of the family doctor program, and official recommendations regarding the use of "green medicine" in primary care stemming from the VI Communist Party Congress (2011). Travel restrictions have been eased, with important repercussions for physicians and scholars working abroad. Contributors to this volume chronicle many of these changes.

The Anthropology of Global Health

Transformations occurring on the island in response to post-Cold War global restructuring have implications for current debates within the anthropology of global health, particularly emergent tensions between national sovereignty and the transnational authority of institutions such as the World Health Organization (WHO) (Lakoff 2010; Crane 2010; Janes and Corbett 2010). As Andaya's chapter illustrates, the priorities of such transnational institutions impact local organization of labor and resources, highlighting some successes and obfuscating disparities that receive less attention on the world stage (e.g. oral and mental health versus maternal and child mortality). Similarly, recent critiques of humanitarian interventions and global health security (Lakoff 2010) resonate with issues raised throughout the contributions to this volume.

Anthropological studies of humanitarian biomedicine have shown that it seeks to alleviate individual suffering, regardless of national boundaries (Lakoff 2010; Fassin 2009), and that the structure of humanitarian intervention tends to take the shape of organizations and philanthropies from advanced industrial countries working on focused projects designed to save lives in the developing world (Lakoff 2010: 66). It follows, then, that these interventions tend to emphasize technical and often temporary fixes such as drugs, vaccines, or bed nets (2010: 67). Julie Feinsilver's seminal volume, *Healing the Masses* (1993), provided the first detailed account of Cuban humanitarianism (medical diplomacy), which is updated in her contribution to this volume. Even a brief exploration of this history begs the question of how consideration of a low-resource country engaging in focused *and* sustained projects troubles current understandings of humanitarian biomedicine.

As detailed in Section Two ("Social accountability and 'the Gift': Internationalism and Cuban Medical Diplomacy") of this volume, Cuban humanitarian intervention, unlike the model replicated by global non-governmental organizations (NGOs) such as the Gates and Clinton Foundations, emphasizes human capital in the form of in-country doctors, training, and scholarships for medical training on the island. Since its initial global health initiatives in the early 1960s in Chile and Algeria, Cuba has had health care professionals working in parts of Africa, Latin America, and Asia (Westhoff et al. 2010).[9] Since that time, Cuba has invested in the training of health personnel from over 108 countries and has established medical schools in eleven (Márquez 2009). The establishment of the Latin American School of Medicine (ELAM) in 1998 after Hurricane Mitch tore through Central America expanded this training, and by 2007 over 3000 foreign doctors had graduated from the school (Westhoff et al. 2010; Huish 2009). Importantly, the training provided is modeled after the process of health care workforce development followed on the island in the 1970s; in addition to the training of doctors and nurses, auxiliary personnel (general, pediatric, health technicians, and obstetric nursing) were also trained to serve in rural areas. Diversification of the health care workforce in this manner has been central to the country's ability to provide free services along the continuum of care, particularly in polyclinics staffed with doctors, nurses, and health technicians (Márquez 2009).

Cuba's involvement in Haiti provides a specific example of how this humanitarian aid has played out. By 2010 Cuba had trained 550 Haitian doctors, and since 1998, almost 7000 Cuban medical personnel had worked in Haiti. In this time, they had given more than 14.6 million consultations, conducted 207,000 surgeries (including 45,000 vision restoration operations through Operation Miracle),[10] attended 103,000 births, and taught literacy to 165,000 (E. Kirk and Kirk 2010). The *Barrio Adentro* (Inside the Neighborhood) program in Venezuela is another example of extended, long-term intervention (Briggs and Mantini-Briggs 2009; Brouwer 2011). However, these efforts have not gone without critique. Cuban doctors working in Venezuela have been heavily criticized (Briggs and Mantini-Briggs 2009), and the investment in foreign aid in the form of human capital has had repercussions on the island. Some have argued that Cuba needs a market for its successful biotechnology products and that their medical humanitarianism is really an attempt to create a dependence upon Cuban medicines in the developing world (Fawthrop

2003; Aitsiselme 2002; Huish and Kirk 2007).

The "care for oil" dilemma posed by Cuba's agreement with Venezuela (Cuba receives 90,000 barrels of oil a day largely in exchange for over 30,000 doctors and medical personnel) coupled with increasing discontent on the island due to the perception that domestic attention is being sacrificed in favor of international solidarity (Alzugaray Treto 2009) are explored in depth in P. Sean Brotherton's contribution to this volume. Ayesha Nibbe expands upon this theme as she reports some of the constraints experienced by those Cubans working abroad (*internacionalistas*) as professors, athletes, and coaches. At the time of her research, overseas Cubans were required to share their earnings with the government. For example, in 2008 the Ministries of Higher Education and Health required that their employees working internationally turn over 50% of their earnings (Eckstein 2010).

Scholars assert that humanitarian interventions, whether pushed through by multinational NGOs or the Cuban government, are largely possible due to the fact that these interventions seek to remain explicitly apolitical. The work of Fassin and others has shown, however, that humanitarian efforts do not stand outside politics, and much can be learned from examination and critique of such efforts (Fassin 2009; McFalls 2010). Alissa Bernstein's chapter explores this dilemma in her depiction of the impacts of the presence of Cuban-trained Bolivian medical students and Cuban doctors on emergent Bolivian health care reform. In a nuanced exploration of what she terms "clinical subjectivity," she shows that despite stated explicit avoidance of political involvement, political, and policy effects, are produced.

Alongside analyses of global humanitarianism, the threat of "emerging infectious diseases" which are seen to threaten wealthy countries, national borders, and nation-state-based systems of public health (Lakoff 2010; Adams et al. 2008) has emerged as a prescient issue in current global health debates. Biosecurity concerns such as these resounded on the island in the 1980s and 1990s as infectious diseases once thought eradicated reemerged, with the reappearance of tuberculosis and hepatitis in the 1990s, the epidemic of neuropathy in 1992,[11] and the dengue epidemic in the 1980s, late 1990s, and early 2000s (Uriarte 2005). In 1997, Santiago de Cuba (Western Cuba) was the first to present a dengue epidemic after the *Aedes aegypti* mosquito eradication. Other smaller outbreaks were reported in Havana in 2000 and 2001, which spurred community-based prevention efforts (Toledo Romani et al. 2007). Charles Briggs explores unintended effects of these efforts in his chapter, an important contribution especially considering the contested space infectious diseases such as dengue occupy within the medical anthropology of the island (Crabb 2001; Hirschfeld 2007) and within Cuban history itself.[12]

An elemental strategy underlying Cuba's response to the health, economic, and diplomatic challenges of the post-Cold War era has been investment in biotechnology. The success and strength of this sector, explored in depth for the first time in Feinsilver's *Healing the Masses* (Feinsilver 1993), underlies Cuba's humanitarian interventions throughout the world, as well as the government's success in treating, and controlling, health threats faced on the island. An example is the Western Scientific Pole, west of Havana, where 38 biotech centers are grouped together.[13] Cuba's first success in this field was the production of interferon in 1981, which was utilized in subsequent years to treat internal bleeding that resulted from dengue fever. Since that time, the island has produced products ranging from vaccines[14] and chemotherapy agents to fetal monitoring equipment, which are exported to over twenty countries including the United Kingdom and Canada. In 2002, the country held over 400 biotechnology patents (Aitsiselme 2002) and had reached 1200 in 2008 (see Feinsilver, this volume). Importantly, Cuba's biotechnology program is integrated into its public health system. Sahra Gibbon's contribution to this volume shows how the advances and possibilities achieved and anticipated in genomics on the island influence and shape emerging shifts in the daily practices of family doctors and structures of polyclinics.

Global reconfigurations in the post-Cold War era have had profound effects in Cuba and for Cuba's presence on the international stage. The chapters that follow, briefly described below, provide rich, detailed explorations of these challenges and their often-unintended effects. As such, contributions to this volume confront, and in some cases highlight, assumed categories and relationships in the anthropology of global health.

Book Organization and Chapters

The book is organized in three thematic sections. The first section, "History, affect, and uncertainty: Public health campaigns and food insecurity," brings together papers that ethnographically explore personal experiences with, and political implications of, two very different kinds of epidemics: an infectious disease (dengue) and food scarcity. Importantly, Briggs' tracing of media reporting of the 2002 dengue epidemic centers on challenges to the control and governance of domestic spaces, while Garth's chapter explores the effects of another form of governance – the ration system – within the context of a dual economy, on identity and mental health. The coping strategies she examines, including searching for ingredients necessary to give food the flavor and meaning that infuse it with "Cubanness," are further explicated in Susannah's Drissi's chapter. Drawing upon central forms of Cuban cultural production (literature and film) and her own experience as a returning immigrant, Drissi describes continuity and change in the meanings of a Cuban *cena* (meal, or dinner), and, like Garth, clearly illustrates how labor informs meaning. Drissi's and Garth's contributions highlight the value of attending to affect, experience, and cultural production – the meanings of food and "a meal" – in the context of food scarcity or insecurity.

The reemergence of dengue in the 1990s and 2000s, as reported in Briggs' contribution, threatened Cuba's legitimacy as a source of medical diplomacy. Without the ability to control the epidemic on the island, how could representatives of the Cuban state internationally be trusted to address the many health challenges they would face? Such concerns, as Briggs details, informed a transformation in the manner in which the Cuban population was constructed in media treatment and reporting on dengue: instead of being celebrated as well instructed biological citizens, they were at fault for the reemergence of this threat and challenged to change their ways. Such a transformation in the treatment of Cuban subjects provides a window into the level of state anxiety about the outbreak and, perhaps, fissures in the socialist body subsequent to the challenges experienced during and since the "Special Period."

The close examination of Cuban medical diplomacy offered in the second section, "Social accountability and 'the Gift': Internationalism and Cuban Medical Diplomacy," offers alternative ways of thinking about humanitarian interventions, as well as the role of history in the forms such interventions take. Beginning with Julie Feinsilver's detailed description of the origins and processes of medical diplomacy since the 1960s, this section draws upon ethnographic research conducted in Cuba, Bolivia and Uganda to illustrate the far reaching and often unanticipated effects of the island's diplomatic efforts. P. Sean Brotherton's rich account of the interplay between medical diplomacy (which removes family doctors from their local communities) and the dual economy (which allows for income differentials and the circulation of hard currency) shows how the history of international aid and investment described in Feinsilver's chapter is experienced on the island. In doing so, he makes visible fissures and rifts not previously analyzed. His introduction of the concept "transactional humanitarianism" for thinking through Cuba's changing relationships and uses of medical internationalism challenges extant histories and highlights the compromises imbricated in Cuba's strategies to adjust to the realities of the post-Cold War era.

The next two chapters take their examination of Cuban medical diplomacy and internationalism off the island to study their effects around the world. Alissa Bernstein follows Bolivian graduates of Cuba's ELAM back to their own country to record how their experience of training in the Cuban model transformed their own subjectivities as physicians and influenced their senses of what it is to provide holistic healthcare in Bolivia. Through in-depth examination of national and personal development in the context of health care reform, she explores the role of Cuban doctors and their students in nation building. Ayesha Nibbe's chapter expands beyond the circulation of medical personnel in the name of Cuban health diplomacy to detail the multiple roles athletes, coaches, and other sports professionals play in the construction and maintenance of the Cuban socialist body on the island and in Africa, specifically Uganda. By tracing the role of sport in the building of the Cuban nation from Fidel Castro's speeches and memoirs to educational policy and democratization of citizen access to stadiums, Nibbe illustrates its role in the exportation of Cuban political and health ideologies.

The last section, "Global, Local, and Personal: Biomedical Practices and Interpretations," brings together three chapters that address gender, sexuality, and personalized medicine in distinct but interlocking ways. Elise Andaya details the daily work involved in prenatal care and monitoring at the neighborhood level. Through her rich ethnographic data, Andaya illustrates the value of "questioning the numbers" and the multiple levels of self-governance involved in the production of globally celebrated statistics in maternal and child mortality. In the process, she shows ways in which patient and provider subjectivities are challenged and transformed. Sahra Gibbon's chapter focuses on her experience working with a team of doctors on the island collecting family history data in order to assess breast cancer risk. As such, she explores the emergence of a new form of personalized medicine in the context of Cuban public health, "community genetics." In her chapter, she unpacks and explicates the cultural context of this particular form of medical practice and its products, and then shows how these practices, as social and cultural products, interact with the local, in this case the tradition of "family medicine," and how they are transformed, through adoption, use, and resistance. Julio César González Pagés' chapter explores notions of masculinity and sexuality on the island, and their implications for thinking about health. Drawing on a rich history and illustrating the entrenched nature of homophobia and discrimination, González Pagés explores changes in cultural conceptualizations of sexuality and normalcy in Cuba. Particularly important is his discussion of the emergence of the concept of *jineterismo* and its implications for rethinking previously held understandings of masculinity in the context of economic and political transformation (see also Hodge 2003; Quiroga 2010; Cespedes 2010).

References

Adams, V., Novotny, T. and Leslie, H. (2008) 'Global health diplomacy', *Medical Anthropology*, 27.4: 315-323.

Aitsiselme, A. (2002) 'Despite U.S. embargo, Cuban biotech booms', North American Congress on Latin America. Online. Available HTTP: <https://nacla.org/article/despite-us-embargo-cuban-biotech-booms> (accessed 15 May 2012).

Alzugaray Treto, C. (2009) 'Continuity and change in Cuba at 50', *Latin American Perspectives*, 36.3: 8-26.

Ayorinde, C. (2004) *Afro-Cuban Religiosity, Revolution, and National Identity*, 1st edn, Florida: University Press of Florida.

Azicri, M. (2009) 'The Castro-Chavez alliance', *Latin American Perspectives*, 36.1: 99-110.

Becker-Barroso, E. (2009) 'For neurologists in Cuba, hope is not embargoed', *The Lancet Neurology*, 8.12: 1088-1089.

Berlanga, J., Fernandez, J.I., Lopez, E., Lopez, P.A., del Rio, A., Valenzuela, C., Baldomero, J., Muzio, V., Raices, M., Silva, R., Acevedo, B.E. and Herrera, L. (2013) 'Heberprot-P: a novel product for treating advanced diabetic foot ulcer', *MEDICC Review*, 15.1: 11–15.

Briggs, C.L., and Mantini-Briggs, C. (2009) 'Confronting health disparities: Latin American social medicine in Venezuela', *American Journal of Public Health*, 99.3: 549-555.

Brotherton, P.S. (2012) *Revolutionary Medicine: Health and the Body in Post-Soviet Cuba*, Durham, NC: Duke University Press.

Brouwer, S. (2011) *Revolutionary Doctors: How Venezuela and Cuba Are Changing the World's Conception of Health Care*, New York: Monthly Review Press.

Brown, D.H. (2003) *Santeria enthroned: art, ritual, and innovation in an Afro-Cuban religion*, 1st edn, Chicago, IL: University of Chicago Press.

Burke, N.J. (2001) 'Creating islands in the desert: place, space, and ritual among Santeria practitioners in Albuquerque, New Mexico', doctoral dissertation, University of New Mexico.

Cabezas, A.L. (2009) *Economies of Desire: Sex and Tourism in Cuba and the Dominican Republic*, Philadelphia, PA: Temple University Press.

Canizares, R.J. (1999) *Cuban Santeria: Walking with the Night*, Destiny Books.

Cave, D. (2011, November 3) 'Cuba to allow buying and selling of property', *The New York Times*. Online. Available HTTP: <http://www.nytimes.com/2011/11/04/world/americas/cubans-can-buy-and-sell-property-government-says.html> (accessed 5 February 2013).

Cespedes, K. (2010) 'Runaway jinateras and addicted pingueros: the narrative crafting of Special Period heroes and deviants', in A. Cabezas, I. Hernandez Torres, S. Johnson and R. Lazo (eds.) *Una Ventana a Cuba y Los Estudios cubanos/A Window into Cuba and Cuban Studies*, San Juan, Puerto Rico: Ediciones Callejon.

Crabb, M.K. (2001) 'Socialism, health and medicine in Cuba: a critical re-appraisal', doctoral dissertation, Emory University.

Crane, J. T. (2010) 'Unequal partners: AIDS, academia, and the rise of global health', *Behemoth: A Journal on Civilization*, 3.3: 78–97.

'Cuba's child mortality was 4.9% in 2011' (2012, January 3) *Prensa Latina*. Online. Available HTTP: <http://cuba-l.unm.edu/?nid=105678&cat=ht> (accessed 16 February 2013).

De Vos, P., Orduñez-García, P., Santos-Peña, M. and Van der Stuyft, P. (2010) 'Public hospital management in times of crisis: lessons learned from Cienfuegos, Cuba (1996–2008)', *Health Policy*, 96.1: 64–71.

Eckstein, S. (2010) 'Remittances and their unintended consequences in Cuba', *World Development*, 38.7: 1047-1055.

Fassin, D. (2009) 'Another politics of life is possible', *Theory, Culture & Society*, 26.5: 44-60.

Fawthrop, T. (2003, November 21) 'Cuba sells its medical expertise', *BBC News*. Online. Available HTTP: < http://news.bbc.co.uk/2/hi/business/3284995.stm> (accessed 13 February 2013).

Feinsilver, J.M. (1993) *Healing the Masses: Cuban Health Politics at Home and Abroad*, Berkeley, CA: University of California Press.

Fernandez, A.D. and Angeles, L. (2009) 'Building better communities: gender and urban regeneration in Cayo Hueso, Habana, Cuba', *Women's Studies International Forum*, 32: 80-88.

Ginsburg, F.D. and Rapp, R. (eds.) (1995) *Conceiving the New World Order: The Global Politics of Reproduction*, Berkeley, CA: University of California Press.

Gosin, M. (2010) '"Other" than black: AfroCubans negotiating identity in the United States', in A. Cabezas, I. Hernandez Torres, S. Johnson and R. Lazo (eds.) *Una Ventana a Cuba y Los Estudios cubanos/A Window into Cuba and Cuban Studies*, San Juan, Puerto Rico: Ediciones Callejon.

Hagedorn, K.J. (2001) *Divine Utterances: The Performance of Afro-Cuban Santeria*, Washington, DC: Smithsonian Books.

Hirschfeld, K. (2007) *Health, Politics, and Revolution in Cuba Since 1898*, New Brunswick NJ: Transaction Publishers.

Hodge, G.D. (2003) 'Colonizing the Cuban body', in A. Chomsky, B. Carr and P.M. Smorkaloff (eds.) *The Cuba Reader: History, Culture, Politics*, Durham, NC and London: Duke University Press.

Huish, R. (2009) 'How Cuba's Latin American School of Medicine challenges the ethics of physician migration', *Social Science & Medicine*, 69.3: 301-304.

Huish, R. and Kirk, J.M. (2007) 'Cuban medical internationalism and the development of the Latin American School of Medicine', *Latin American Perspectives*, 34.6: 77-92.

Janes, C. and Corbett, K. (2009) 'Anthropology and global health', *Annual Review of Anthropology*, 38.1: 167-183.

— (2010) 'Anthropology and global health', in B. Good, M.M.J. Fischer, S.S. Willen and M.J. DelVecchio Good (eds.) *A Reader in Medical Anthropology: Theoretical Trajectories, Emergent Realities*, New Jersey: Wiley-Blackwell Publishing, Ltd.

Kirk, E.J. and Kirk, J.M. (2010) 'Cuban medical cooperation in Haiti: one of the world's best-kept secrets', *Cuban Studies*, 41: 166-172.

Kirk, J.M. and Erisman, H.M. (2009) *Cuban Medical Internationalism: Origins, Evolution, and Goals*, 1st edn, United Kingdom: Palgrave Macmillan.

Lakoff, A. (2010) 'Two regimes of global health', *Humanity: An International Journal of Human Rights, Humanitarianism, and Development*, 1.1: 59-79.

Latin American Institute and Centro de Estudios Sobre America (LAI & CEA) (1995) 'Migración Cubana: crisis de los balseros en el verano de 1994', *Cuba en el Mes*, Dossier 7.

Maal-Bared, R. (2006) 'Comparing environmental issues in Cuba before and after the Special Period: balancing sustainable development and survival', *Environment International*, 32.3: 349-358.

Márquez, M. (2009) 'Health-workforce development in the Cuban health system', *The Lancet*, 374.9701: 1574-1575.

Mason, M.A. (2002) *Living Santería: Rituals and Experiences in an Afro-Cuban Religion*, Washington, DC: Smithsonian Books.

Mazzei, J. (2012) 'Negotiating domestic socialism with global capitalism: so-called tourist apartheid in Cuba', *Communist and Post-Communist Studies*, 45.1-2: 91-103.

McFalls, L. (2010) 'Benevolent Dictatorship: The Formal Logic of Humanitarian Government', in D. Fassin and M. Pandolfi (eds.) *Contemporary States of Emergency: The Politics of Military and Humanitarian Intervention*, New York: Zed Books.

McNeil, D.G. (2012, May 7). 'A regime's tight grip: lessons from Cuba in AIDS control', *The New York Times*. Online. Available HTTP <http://www.nytimes.com/2012/05/08/health/a-regimes-tight-grip-lessons-from-cuba-in-aids-control.html> (accessed 13 February 2013).

Quiroga, J. (2010) 'Cuba: la desaparición de la homosexualidad', in A. Cabezas, I. Hernandez Torres, S. Johnson and R. Lazo (eds.) *Una Ventana a Cuba y Los Estudios cubanos/A Window into Cuba and Cuban Studies*, San Juan, Puerto Rico: Ediciones Callejon.

Reed, G. (2012) 'Revolutionizing gender: Mariela Castro MS, Director, National Sex Education Center, Cuba', *MEDICC Review*, 14.2: 6-9.

Rodríguez, A. (2010, January 20) 'Cuba performing state-sponsored sex change surgery', *Huffington Post*. Online. Available HTTP: <http://www.huffingtonpost.com/2010/01/19/cuba-performing-statespon_n_428681.html> (accessed 5 February 2013).

Roland, L.K. (2010) *Cuban Color in Tourism and La Lucha: An Ethnography of Racial Meaning*, Oxford: Oxford University Press.

Tancer, R.S. (1995) 'The pharmaceutical Industry in Cuba', *Clinical Therapeutics*, 17.4: 791-798.

Toledo Romani, M.E., Vanlerberghe, V., Perez, D. Lefevre, P. Ceballos, E. Bandera, D. Baly Gil, A. and Van der Stuyft, P. (2007) 'Achieving sustainability of community-based dengue control in Santiago de Cuba', *Social Science & Medicine*, 64.4: 976-988.

Uriarte, M. (2005) 'Social policy responses to Cuba's economic crisis of the 1990s', *Cuban Studies*, 35: 105.

Westhoff, W.W., Rodriguez, R., Cousins, C. and McDermott, R. J. (2010) 'Cuban healthcare providers in Venezuela: a case study', *Public Health*, 124.9: 519-524.

Whiteford, L.M. and Branch, L.G. (2009) *Primary Health Care in Cuba: The Other Revolution*, Lanham, MD: Rowman & Littlefield Publishers.

Endnotes

1   *Periodo Especial en Tiempos de Paz.*

2   The AfroCuban religion Regla de Ocha is also popularly known as Santería (See also Burke 2001; Ayorinde 2004; Brown 2003; Canizares 1999; Hagedorn 2001; Mason 2002).

3   These laws were enacted to deny U.S. multinationals the right to trade with Cuba through third countries, and to deny countries that traded with Cuba U.S. aid. The embargo against Cuba, in place since 1962, prohibited direct United States-Cuba trade (Eckstein 2010).

4   The minimum wage was more than doubled in 2005. At the same time, pension and welfare payments were increased. To alleviate the impact of costs associated with these changes, the retirement age was raised by five years in 2008 (Eckstein 2010).

5   The dollarization strategy was transformed in 2004 when the government transitioned from U.S. dollars and other foreign currency to *pesos convertibles* or CUCs – promissory notes with dollar value on the island. This change officially restricted the informal circulation of dollars (while dollars still remained legal) and permitted the state, through the imposition of a 20% surcharge on conversion, to tax remittance recipients and those with access to hard currency via the black market and tourism (Eckstein 2010). In addition to higher prices charged in state-run stores and taxes for remittances received via wire services, the 20% surcharge effectively channeled remittances, which account for 80% of hard currency that Cubans attain, into state coffers (Eckstein 2010). This 20% charge was lowered to 13% in November 2011.

6   Shops selling religious supplies (herbs, pottery, beads, etc.) for practitioners of Afro-Cuban religions.

7   In April 2009 U.S. President Barak Obama removed remittance-sending restrictions. Prior to 2009, Cubans largely avoided the restrictions by sending money via informal transfers (e.g. with people traveling to the island, often for a fee).

8   In 2008 the figures reached 4.7, in 2009, 4.8, and in 2010, 4.5% ('Cuba's Child Mortality' 2012).

9   Some argue this practice of internationalism is the source of the genetic diversity of Cuba's HIV/AIDS epidemic, which features 21 different strains of the virus (McNeil 2012).

10 Cuba's Operation Miracle began to offer free eye surgery for cataracts, glaucoma, diabetes, and other vision problems to people unable to afford it in

2004. A joint program established between Havana and Caracas has been flying Venezuelans to the island cost-free for surgeries since that time. Within a few years, 28 countries of Latin America and the Caribbean were participating in the program, and operations restoring patients' sight number close to 850,000. By early 2007, 13 eye clinics in Venezuela were performing thousands of operations (Azicri 2009).

11   In the early 1990s an optic neuropathy epidemic linked to malnutrition affected more than 50,000 people in Cuba. At its peak, there were over 3000 cases per week, but this number declined after the Ministry of Public Health implemented a country-wide campaign of multi-vitamin supplementation (Becker-Barroso 2009).

12   All Cuban school children learn the story of Dr. Carlos Finlay and the mosquito, a source of national anger and frustration for what is described as his mistreatment at the hands of history. Carlos Finlay was the first to theorize that a mosquito was the vector by which yellow fever was transmitted, arguably one of the most important advances in tropical medicine. His theory was discounted for years, despite increasing evidence and the seriousness of the yellow fever epidemic in Havana and New Orleans at the time. On August 14, 1881, Dr. Finlay presented a paper entitled "The Mosquito Hypothetically Considered as an Agent in the Transmission of Yellow Fever Poison" at the Academy of Medical, Physical, and Natural Sciences of Habana. In this presentation, Dr. Finlay stated that the *Culex fasciatus* mosquito, today *Aedes aegypti*, was the vector of the disease, and suggested that employing strategies to prevent the development of the *Culex* mosquito could prevent yellow fever (these strategies are the basis for those discussed in Briggs' analysis of the 2000 dengue epidemic, this volume). Finlay was ridiculed for years, both on and off the island, as attention was focused on poor sanitary conditions as the culprit. Credit for his discovery was eventually taken by Major Walter Reed, head of the fourth Yellow Fever Commission sent to Havana by the U.S. Government in 1900. Reed, through a variety of experiments, validated Finlay's theory and went on to credit him in private correspondence for his achievements, but continues to be listed in official sources as the discoverer of the mosquito vector. The lack of recognition for Finlay (he was nominated seven times for the Nobel Prize, but never received the award), the ridicule he endured, and having to have his findings "confirmed" by a non-Cuban has rankled Cuban nationalism, before and after the Revolution. This contentious history is ever present in the consciousness of the well informed Cuban population described by Briggs, and is a reference point for the reemergence of dengue, also transmitted via the *Aedes aegypti*.

13   The Center for Genetic Engineering and Biotechnology (CIGB), the most prominent of Cuba's biotechnology centers, was established in 1986 in the Western Scientific Pole. There are 14 scientific poles on the island.

14   Cuba's hepatitis B vaccine is on the World Health Organization (WHO) list of vaccines purchased by United Nations agencies (Aitsiselme 2002), and their antimeningococcal type B vaccine – directed toward a form of the disease that causes significant deaths in infants – has been exported around the world. According to a 2002 report, this vaccine caught the attention of the pharmaceutical company SmithKline Beecham (now Glaxo SmithKline), which subsequently reached an agreement with the Carlos Finlay Institute in Havana to market the vaccine globally. Despite limitations on trade imposed by the Helms-Burton Act, SmithKline Beecham was able to demonstrate the unique value of the Cuban vaccine and received a license from the U.S. Treasury Department to bring the vaccine to the U.S. market (provided the vaccines were produced in SmithKline Beecham facilities) (Aitsiselme 2002).

Section I

History, Affect, and Materiality:
Public Health Campaigns and Food Insecurity

1

The Biocommunicable State: Dengue, Media, and Indiscipline in La Habana[1]

Charles L. Briggs

Medical anthropologists have struggled in recent years to create narratives that do not fall into predetermined binaries or reproduce biopolitical subjects and objects whose definitions and contours have been determined in advance. But writing about Cuba presents particularly thorny problems in this regard. The politics of Cuban exceptionalism looms equally large in functionalist accounts that celebrate all features of Cuban socialism as unique departures from the evils of neoliberalism and self-interest in the organization of health care as well as those that condemn socialist medicine as the embodiment of political subjugation and substandard medical care. Here outdated Cold War political dualities seem multiplied by mechanistic logics and linear narratives of modernity that have long inhabited medical domains, even as medical anthropologists have provided some of the most effective critiques and explored alternative conceptual and narrative frameworks elsewhere. Both types of narratives embody what I have referred to as bioexceptionalism (Briggs 2011), the notion that Cuba systematically departs from broader patterns on the basis of health-related characteristics and defies the material constraints that would seem to determine biomedical infrastructures and health outcomes.

Given the way that the Special Period has extended the deep biopolitical significance of medical research and health care even while straining firmly established health infrastructures, one might think that it would have pushed scholars to move beyond binaries and functionalist logics. Nevertheless, neoliberal logics seem to reemerge in both accounts of heroic struggles in which revolutionary ideology and praxis have enabled Cubans to maintain health infrastructures and reject self-interest and material constraints *and* those of the emergence of Adam Smithian propensities to truck and barter, perspectives that locate practitioners and patients alike in biomedical black markets.

In this chapter I explore the 2002 dengue fever epidemic in La Habana. Accounts of dengue epidemics have appeared in scholarly analyses as characteristic embodiments of Cuban health ideologies, infrastructures, professional practice, and popular participation. Dengue would seem to challenge linear narratives due to its cyclic character. The noise created by the buzzing, breeding, and bites of non-human actors, the *Aedes aegypti* mosquitoes that "transmit" dengue, would similarly appear to complicate efforts to construct totalizing stories in which politics performatively and directly engender medical subjects, objects, and practices or vice-versa. Nevertheless, rather than complicating received narratives of politics and health in Cuba, analyses of dengue epidemics often seem to confirm their basic contours.

Political scientist Julie Feinsilver's comprehensive *Healing the Masses* suggests that the 1981 epidemic "epitomizes the intersection of symbolic politics and health through the ensuing symbolic war between Cuba and the United States" (1993: 86). She argues that the dengue "campaign" quintessentially embodied Cubans' "national capacity to mobilize," creating a "massive effort" in which "everyone participated" (1993: 88). While noting the human and economic "costs" of the epidemic, she argues that, "No government in the Third World and few in the developed countries could have achieved as much as rapidly as the Cubans did, because most lack this national capacity to mobilize" (1993: 88). Feinsilver suggests that the "campaign" demonstrates that health infrastructures, popular participation, and health education "can be used successfully for larger, symbolic political ends (1993: 90), both "enhancing communal bonds" (1993: 89) and generating symbolic political capital.

Katherine Hirschfeld's first-person account of her experience as a patient in the 1997 dengue epidemic in Santiago could not be more dissimilar. Dengue similarly provides a synecdoche that demonstrates the tenor of Cuban biopolitics, but here the diagnosis is one of dysfunction, disinformation, and health care as political suppression. Health professionals and patients alike actively suppress overt references to dengue, referring to it euphemistically as "a virus." Every aspect of official responses to the epidemic provides evidence of the failure of Cuban health infrastructures and medical practices and the seeming indifference – and often hostility – of health professionals. Fellow patients do not give voice to revolutionary and collective constructions of health care but to its failure, both in mocking and bitter remarks on infrastructures and assessments of the lack of care on the part of professionals: "'You have to yell to get their attention, otherwise they'll ignore you'" (Hirschfeld 2007:

63). Rather than providing evidence for communal solidarity and the medical and symbolic success of Cuban capacities for mobilization, Hirschfeld's stay in a Santiago hospital made her feel "like nobody at all – just another invisible figure, held incommunicado with a politically incorrect case of 'virus'" (2007: 60-61).

P. Sean Brotherton's recent *Revolutionary Medicine* devotes considerable space to developing an ethnographic picture of the 2002 dengue epidemic in La Habana. Brotherton was also interpellated by the Cuban State through dengue, not as a patient but through his futile attempts to dissuade fumigators from spraying his apartment – which led to a visit by "a middle-aged man dressed in typical military garb" who "firmly told me that I had no choice in the matter. 'Fumigation is the law', he stated" (2012: 127). Brotherton points to a surplus of discourse about dengue, as was apparent in 1981, and he echoes Feinsilver's analysis of dengue mobilization as "a form of collectivization" (2012: 131) and "as reflecting the symbolic importance of Cuba's battle against the United States" (2012: 129). He cites neighbors who read apparent contradictions between intense, systematic discursive and material state responses to dengue and decaying infrastructures and popular indifference as proving that "'things in this country do not have any logic!'" (2012: 127). Rather than following Hirschfeld in suggesting that such comments demonstrate Cuban socialism's collapse, however, Brotherton suggests that physicians and non-professionals alike strategically seize on dengue discourse to promote their own interests in cleaning up their neighborhood. Analysis of the dengue epidemic embodies his larger thesis regarding the effects of the Special Period on state policies and their effects on individual lives, converting revolutionary actors who embrace socialist ideologies and rely on state institutions into self-interested individuals. Brotherton identifies two new types of Cuban medical subjects, those who draw on access to foreign currency in becoming "active health consumers" and others who rely on "informal" networks centering on black market activities, small for-profit businesses, and money from tourists obtained through hustling and prostitution (2012: 7).

These narratives present important differences, reading dengue mobilization as demonstrating the strength of the Cuban revolution (Feinsilver), its failure (Hirschfeld), or as the emergence of new strategies for surviving its contemporary transformation (Brotherton). Beyond similarly pointing to the centrality of militarization in dengue responses, they seem to have two central characteristics in common. First, all three authors seem to see dengue epidemics and Cuban responses as embodying consistent relationships between health infrastructures, subjectivities, and practices, whether of political will and communal solidarity (Feinsilver), active state oppression and the passive complicity of health professionals and patients (Hirschfeld), or strategic participation in order to promote forms of sanitation that have declined in the face of decaying socialist resources and the erosion of communal participation through the rise of self-interest (Brotherton). Second, dengue enables these scholars to jump scales, constituting a synecdoche that helps them move between accounts of individual perspectives, observations of particular clinical settings or home visits by *brigadistas* (members of health brigades) and others on the one hand and accounts of municipal and national policies, practices, and discourses on the other.

In short, whatever narrative you construct of health and socialism in Cuba, dengue confirms it. My goal here is not to challenge these accounts of the epidemics of 1981, 1997, and 2002 but rather to explore how dengue can disrupt and complicate linear, teleological narratives of health in Cuba. I am particularly interested in pointing to how the 2002 epidemic revealed disjunctures between infrastructures, subjectivities, and practices.

As many observers have pointed out, health was crucial to fashioning Cuba as a global medical power and providing a continuing symbol of the fruits of the revolution – and thus symbolizing what Cuba would have to lose if socialism collapsed. Over four decades of health education, access to health care, massive public health campaigns, extensive coverage in print, radio, and television media, and the health-focused website Infomed reportedly produced a population that was projected as the most medically literate in the world (Díaz Briquets 1983; Feinsilver 1993; Whiteford and Branch 2009).

Until it all seemed to disappear overnight in January of 2002. After more than 15 years without cases, dengue was again reported in Cuba in 1997. Returning in 2001, a large outbreak emerged in La Habana itself in 2002, prompting a massive mobilization. Speaking not only as *Comandante* of the Revolution but also as symbolic head of the "battle" or "war" against dengue, Fidel Castro[2] himself seemed to declare that Cuban bioexceptionalism had died or at least was gravely ill. He argued that the 1997 outbreak demonstrated that "evidently there was carelessness when no cases appeared for 16 years" (Núñez Betancourt 2002b). He pointed to grave weaknesses in the discipline, organization, and epidemiological surveillance in the health system and a failure of health education (Núñez Betancourt 2002h). Castro admitted that dengue had not been eliminated in over a year, resulting in an "unfavorable situation" that produced "a loss of confidence on the part of the population." He read quotes from "residents of various districts that reveal a certain discontent with the work that was being done" (Núñez Betancourt 2002b). More broadly, news coverage of dengue, the arena in which reporters seemed to most explicitly accept the subordination of their own professional ideology to state directives, as I suggest below, ironically constituted a case in which "the state" seems to totalize itself, "biomedicine," and "the people" precisely in such a way as to project their individual and collective *failure*.

Ouch! This outcome would seem to call into question the truths that Cuban studies hold to be self-evident: that health is presented as the epitome of the successes of the revolution, that the press has minimized awareness of hardships and vulnerabilities during the Special Period, and the media – especially the three leading print "organs" of the Revolution – are direct reflections of state power. And the Cuban people, far from being exalted for their medical literacy and responsiveness to the advice of physicians, were accused of indiscipline, disorganization, ignorance, and deafness. Castro famously called the epidemic an act of U.S. bioterrorism, suggesting that it was deliberately spread as an attack on the Cuban state. Whether or not his epidemic origin narrative resonated, a mosquito and its viral cargo seemed to have placed the revolution on the line. The emergence of cases in the heart of La Habana seemed to mock state power and jeopardize the spatial locus of Cuban modernity. A resident complained in *Granma* – the official "organ" of the Cuban Community Party – that "[t]he presence of mosquitoes at sundown is shocking ... it is not logical that in the center of the capital we should have to sleep with mosquito netting" (Núñez Betancourt 2002d).

Biocommunicability and the Performative Work of Health News

A central concern of medical anthropology, partly inspired by the work of Michel Foucault (1976[1978], 1997), has been with how medical subjectivities are produced and the complex ways they interpellate individuals and populations. Many of us have emphasized that medical subjectivities are multiple, such that seemingly contradictory forms coexist within individuals, sites, texts, and technologies. One of my principal foci in recent years has been cultural models that project the production, circulation, and reception of knowledge about health, which I refer to as biocommunicable models (Briggs 2005). These models, which are always multiple and sometimes competing, create medical subjects that are often hierarchically organized into privileged producers (such as medical researchers and epidemiologists), less prestigious but still professionally sanctioned circulators (such as health educators and promoters), persons whose roles are restricted to receiving knowledge and embodying it (laypersons who are recognized as biocommunicable citizens), and individuals and populations deemed to simply be out of the loop (racialized populations are often stereotypically thrust into classes of resistant, ignorant, or noncompliant subjects). Biocommunicable models generally purport to be simple reflections of actually existing knowledge-making practices; nevertheless, they are simplifications and projections of processes whose complex, heterogeneous, and shifting characteristics do not conform to simple and consistent patterns. Nevertheless, rather than simply being abstract idealizations that exist apart from the "real" facts of how we produce and receive medical knowledge, the subjectivities they produce have social lives with forms of affective and bodily presence that invite particular forms of knowing and marginalize others. When interpellated as patients, we are invited to trust and respect our physician and asked to inscribe her/his words on our bodies or scolded for noncompliance, ignorance, resistance, or self-medication - that is, for believing that we can produce knowledge about our bodies, select which dimensions of medical advice we wish to assimilate, and decide how we wish to embody it.

Biocommunicable models emerge in a wide range of sites, each of which is associated with distinct discursive modes of representing them. Communicable models embedded in articles in *The Lancet* or the *New England Journal of Medicine,* for example, are seldom made explicit - simply presupposing an evidence-based production of knowledge, its circulation via peer review and National Institute of Health regulated professional practices, and its reception by biomedical and other scientific professionals. (The distance between implicit communicable models and the pragmatics of discourse sometimes emerges when it becomes evident that articles on pharmaceuticals have been drafted by medical writers at the behest of public relations and marketing personnel, possibly unbeknownst to their putative authors.) In health education and promotion, however, models generally are rendered explicit, specifying the locus of knowledge production in biomedical domains and how lay audiences are expected to interpret and embody them.

In collaboration with Daniel Hallin, Clara Mantini-Briggs, and others, I have been particularly interested of late in one of the sites in which biocommunicable models are most widely disseminated and become particularly visible - news coverage of health issues (Briggs and Hallin 2007). It was precisely the proliferation of health news in Cuba, along with the special positionality of the media vis-à-vis the state, that brought me to the island. In a previous article (Briggs 2011), I analyzed how the production, circulation, and reception of health knowledge is culturally modeled in Cuban radio, television, and newspaper coverage, and I documented the perspectives of health and media professionals and laypersons on the mediatization of health. I found two dominant modalities. One, also evident in media coverage in capitalist countries, views the role of health news as tracking the production of health knowledge in professional institutions (laboratories, leading hospitals, etc.); its circulation by clinicians, health educators, and journalists; and its reception by laypersons (whose role is to passively assimilate professional knowledge). A second, more directly tied to the Cuban experience, reports *logros de la revolución* ("achievements of the revolution"). Here laypersons are projected as spectators, not so much receiving knowledge as enjoying its fruits in biomedical services. *Logros* stories contain a fascinating contradiction regarding how national subjects are interpellated. On the one hand, laypersons are positioned on the outside, looking inside health institutions as the latter embody scientific communication; laypersons largely enter as beneficiaries of high-quality medical services. On the other hand, all Cubans are affectively projected in biocommunicable terms as receiving the news with similar feelings of solidarity and pride.

In interviews, media and health professionals constructed laypersons in distinct ways – as embodying Cuban bioexceptionalism through heightened bioliteracy and responsiveness to medical advice, or as innocent, gullible, and prone to circulating faux information – and in need of close supervision. Interviews and observations conducted between 2003 and 2008 suggested that some Cubans who are not media or health professionals were not particularly interested in health news. Many, however, were so focused on building a broad archive of health knowledge that they deemed themselves to be "frustrated doctors" – as being as knowledgeable as their physicians. Instead of focusing on features relevant to their "risk factors" or individual health aspirations, a strong trend in the United States (Briggs and Hallin 2007), many laypersons told me that they want to learn everything about everything medical. Some suggested that their knowledge base was broader than that of their physicians, who were overly specialized and less familiar with "green medicine," which was contested by many physicians in the 1990s, and with the emergence of complementary and alternative medicine. Rather than citing these responses as demonstrating the success of the circulation of health knowledge in Cuba, however, many media and health professionals criticized what they saw as a problematic lay claim that "all Cubans are doctors!" Many associated overly active reception of health knowledge with the reportedly equally widespread problem of self-medication. Note the irony here: hyper-identification with the biocommunicable subject-position of lay-receiver confers both the status of idealized Cuban citizenship *and* arrogant claims to medical agency!

What, then, are we to make of January 2002? The Cuban press has often been criticized for turning a blind eye to effects of the Special Period. Although Castro continued to refer to the Special Period as ongoing in the 2000s, the government had reestablished more control over public discourse and the economy by the end of the 1990s (Hernández-Reguant 2008). Lauding the "achievements of the revolution" served a crucial political function in drawing attention away from continuing shortages of medicine and services and toward the survival of health infrastructures as major benefits of the revolution. Thus, why would the official media envision the health system in January 2002 as having significantly failed the Cuban people and lost their confidence? If biomedical literacy was a major feature of Cuban bioexceptionalism, why portray laypersons as either not getting the message or actively resisting it? The crisis was short-lived: less than three months later, Castro declared that dengue was eradicated and the *Aedes aegypti* vector controlled, and the *logros* model burst forth again as a vehicle for celebrating the revolution's success in overcoming two more enemies – a mosquito and its viral passenger.

In this chapter I focus on this period of rupture in which clinical medicine, epidemiology, and health communication seemed to be falling apart. As Latour (1987) has suggested, scholars might learn more by examining points at which black boxes are open than those in which their closure permits business as usual. The epidemiological features of dengue – the fact that domestic spaces provide crucial breeding grounds for *Aedes aegypti* and sites of transmission for dengue - make the biomedical focus of the crisis particularly instructive (Gubler and Kuno 1997; Nichter 2008; Rigau-Pérez et al. 1998; Whiteford and Hill 2005). Official anxiety over perceived communicative failures points to the role of the mass media in fashioning health into a major symbol of the revolution and means of projecting Cuba as "a world medical power" (Feinsilver, 1993: 90; Eckstein, 1994). These apparent exceptions to bioexceptionalism did not burst forth *de novo* in January in 2002 but rather form part of a long history of state critiques of indiscipline and a lack of revolutionary spirit. We could thus view them as consistent with long-standing revolutionary ideologies and practices. I wish, however, to distance myself from this approach just as much as from viewing them as proof of the collapse of socialism or the prevalence of individual strategizing. Rather, I wish to reflect on how dengue coverage might assist us in producing a different kind of narrative about health in Cuba, one that focuses on ruptures and specificities. We might thereby look more to contingent correspondences between revolutionary ideologies and infrastructures and the uncertain and not entirely predictable ways that are embodied. I may not exactly ask, with Timothy Mitchell (2002), can the mosquito speak? But I am interested in thinking a bit more about what all that buzzing, with its massive media amplification, can tell us about how constructs such as revolution, state, and people are being produced in Cuba.

Dengue in La Habana, January 2002

The history of dengue in Cuba has been widely discussed (see Kourí et al.1986, 1998; Guzman and Kourí 2002; Guzmán et al. 2006; González et al. 2008; Whiteford and Hill 2005). A number of features of dengue epidemiology are directly relevant to issues of health communication and popular participation. Since even a single flowerpot can become a major site of reproduction of the *Aedes aegypti* mosquito that "transmits" dengue, control efforts seek to eliminate household breeding sites – often involving direct surveillance conducted by health workers. Households that fail to "comply" can be identified by neighbors and health workers as "threats" to their neighbors. Second, dengue, which can be confused with a bad case of the flu, does not ordinarily require hospitalization, and treatment of less severe cases often consists of bed rest and control of symptoms. Nevertheless, dengue symptoms can become quite serious and require hospitalization to increase survival rates.

Anderson (2006), Arnold (1993), Mitchell (2002), and others have fruitfully analyzed how the mobility of mosquitoes, like efforts to control them and the diseases for which they become "vectors," are deeply inscribed in histories of modernity; this is no less true of dengue than of malaria. If that symbol of chemical modernity, DDT, helped bring development and progress to the American tropics, as embodied in the looming figure of Venezuelan Arnoldo Gabaldón, accounts of dengue epidemiology often see restrictions on DDT use as responsible for dengue's reemergence, along with that epitome of late modernity – global travel and transportation. Dengue cases were reported in 1977 in Cuba. A more lethal serotype (DEN-2) appeared there in 1981, resulting in the first outbreak of dengue hemorrhagic fever in the Americas. Morbidity was high, but only 158 deaths occurred. Dengue was eradicated in only three months, thanks to a massive mobilization of personnel, house-to-house surveillance, spraying, sanitary laws aimed at the disposal of containers and other waste, extensive hospitalization, and massive health education efforts (Feinsilver 1993; Gubler and Kuno 1997; Kourí et al. 1986). Another outbreak in 1997 was controlled by 1998 (Kourí et al. 1998).

Given the centrality of bioexceptionalism to Cuban national narratives, becoming the epicenter of a vector-borne "tropical disease" was as politically sensitive as public health problems get. Add the focus on domestic spaces and the possible need for state intervention into household aesthetic, hygienic, and bodily micro-politics, and you have the making of a precarious and fascinating convergence of non-human actors, sites of social and biological reproduction, and a most anxious state.

My analysis of news coverage of the 2002 dengue epidemic forms part of a larger study of news coverage in Cuba, itself a component of a collaborative project also carried out in Argentina, Ecuador, Mexico, the United States, and Venezuela. In Cuba I collected all health-related articles appearing in 2002 in the three leading national newspapers, *Granma, Juventud Rebelde*, and *Trabajadores*, totaling 807 articles.[3] Dengue was the focus of 148 of them, including 40 in one month alone (March) in the official "organ" of the Cuban Communist Party, *Granma*. I have also surveyed health coverage from 2003 through the present, picking weeks at random for systematic sampling, and sampled health news in regional outlets. I conducted interviews during fieldwork periods between 2003 and 2008. Ethnographic observation focused on media institutions, including script development, editorial and production decisions, and reporting; I observed everyday work and meetings in clinical and public health institutions; and I spent time in Cuban homes, including observing media reception. All interviews took place in La Habana, but some participants were from other regions.[4]

Hirschfeld (2007) and Brotherton (2012) usefully make frequent reference to media coverage as a means of informing their ethnographic accounts, respectively, of the 1997 and 2002 dengue epidemics. Media analysis is largely placed, as is the case in most medical anthropology, outside an ethnographic focus. Although my analysis is informed by work conducted in June 2003 with one of the dengue control teams, I reverse this relationship here, viewing dengue control through media analysis, which is itself placed within an ethnographic lens.

Mediatizing an Epidemic Battle

Cuban health news reporting is clearly shaped by the configuration of health and media institutions. Like *Granma*'s relationship to the Cuban Communist Party, *Trabajadores* is affiliated with the Cuban Workers Confederation and *Juventud Rebelde* with the Young Communist League. My interviewees never claimed that their reporting was unconstrained; they never pretended to be unaffected by health ministry agendas. Reporters hold periodic meetings where high officials detail research projects, plans, new infrastructures, problems, and goals; journalists insist, however, that officials seldom suggest stories or shape agendas. Journalists positioned themselves as "trying to find a balance" in researching government programs and consulting leading specialists as well as gauging audience concerns in generating story ideas and researching them.

One of the things that made dengue coverage so interesting is that here reporters seemed to *over*estimate their subservience to state agendas. *Trabajadores* reporter Josefina Martínez distinguished "topics, like the dengue campaign, that are imposed on us" from what they referred to as "journalism proper." Unlike the rarity with which other subjects are reportedly "imposed," Martínez suggested that "they make us talk about dengue not just when there is an outbreak somewhere in the country - it's a theme that we more or less try to maintain." Nevertheless, dengue coverage dropped shortly after the epidemic ended as journalists shifted to other foci. The degree of coordination of dengue coverage during the 2002 epidemic was signaled by the use of the same logo, featuring an *Aedes aegypti* mosquito in a crosshairs with the words "Offensive against the Enemy," on each page in which a dengue article appeared in *Granma, Juventud Rebelde,* and *Trabajadores* [see Figure 1]. An assessment of the epidemic authored by scientists from the Pedro Kourí Institute of Tropical Medicine (IPK) and other institutions cited "the participation of

Figure 1. *Aedes aegypti* mosquito in the crosshairs

the mass media under a single command (*dirección única de mando*)," along with popular participation and "strong political will" as one of five contributions to the dengue response and the "lessons learned" by controlling it (Guzmán et al 2006: 288).

Some dengue articles followed what Hallin and I refer to as the doctor's orders model for projecting the production, circulation, and reception of knowledge about health. An article that appeared in *Granma* provides a good example. For over a quarter century, *Granma* has featured a weekly feature called "*Consulta Médica*" ("A Visit to the Doctor's Office"), in which a journalist whom I will call Francisco García interviews a leading specialist. García suggested to me that he uses a question-answer format in these articles in order to satisfy his physician-readers that he "is maintaining the seriousness and rigor associated with a medical subject." This format thus uses a textual convention to frame one voice as the embodiment of the communicable subject-position of knowledge-producer; by reproducing questions in his own voice, he projects the process of transmitting knowledge to lay receivers.

A 22 February 2002 article entitled "Offensive Against the *Aedes Aegypti* Mosquito" centers on Juan Camacho Cabrera, a "specialist of the Regional Center for Hygiene and Epidemiology of Marianao." Camacho details the use of abate, an organophosphate larvidice used to prevent the reproduction of *Aedes aegypti*, whose use has been questioned by medical anthropologists (Khun and Manderson 2007). García leads with a classic premise of doctor's orders stories, a construction of their designated receivers as laypersons who are ignorant and therefore in need of the knowledge it proffers: "A survey just conducted in eleven districts of the capital produced the result that the uses of abate are not known...." Camacho then seems to take responsibility, on the part of public health institutions, for this projected popular ignorance: "I think that perhaps we have been lacking to some extent the necessary educational element to transmit the ideas that we must have about abate...."

Camacho's statement presents a classic assertion of linear, hierarchically ordered biocommunicable models. According to this type of projection, medical professionals decide what laypersons should know and bear responsibility for "transmitting" it to them, while the latter "must" assimilate biomedical knowledge. García also reports conversations with a toxicologist and refers readers' queries to a professor of public health in providing "an official scientific opinion" about the effectiveness of abate in larvae control and its innocuousness for humans, animals, and the environment. Although laypersons are cast as largely ignorant of abate use, García suggest how reception of biomedical knowledge can be achieved, giving one man a "score of 100" on his knowledge of abate. The model citizen explains, however, that his wife "is a nurse and I know those details by heart." García's article was similar to many stories in *Granma* and *Trabajadores* and numerous television and radio programs that reproduced scientific discourse about dengue, its symptoms, *Aedes aegypti* reproduction, past epidemics, and dengue research conducted by IPK and other leading Cuban biomedical institutions.

Other dengue stories adopted the *logros* format. On 12 January, early in the anti-dengue "battle," *Granma* reporter Núñez Betancourt described a meeting in the symbolic site of the Karl Marx Theater, presided over by Castro, that included "forces that will launch the battle against the mosquito in the capital." Evoking a nationalist totality, Núñez Betancourt (2002b) quoted Castro's exhortative message: "the country has a colossal force," and all Cubans are combatants who must fight decisively and without resting. Núñez Betancourt evoked the language of the statistical and institutional variant of the *logros* model, outlining the composition of the 10,737-person "force" that will combat dengue and the institutions from which they were drawn as well as the special contingent of family physicians and epidemiologists from other provinces who would be working in La Habana. Castro then referenced the trucks, tractors, "Bazuca" fumigators, and other equipment that would be used, and how Cuban factories were producing lids for water storage vessels that would be massively distributed. Linking material and human aspects, he suggested that "the forces that this Revolution can mobilize are incalculable" (Núñez Betancourt 2002b). Castro announced that the goal was both to control dengue and to demonstrate the strength of the revolution.

Nevertheless, dengue seemed to have turned *logros* into a failed logos, not touting successes but contrasting the precarious present with a hopefully glorious future. In his initial intervention in Karl Marx Theater, Castro boldly stated that the battle *must* be successful. "Men and women, revolutionary Cubans, we're going to show friend and enemy alike what is a true revolution, what is socialism" (Núñez Betancourt 2002b). The article ended on a classic note of triumphalist bioexceptionalism: "if for many countries in many aspects the name for hope is Cuba, the solution that we find for controlling dengue, beyond saving the lives of children, adolescents, and adults, will increase what people expect from our country." Cuban medical diplomacy, the role of medicine in projecting Cuba's international presence, was seemingly on the line with each mosquito and virus reproduced on the island.

An Anxious State: Projecting Biocommunicable Failure

Both doctor's orders and *logros* narratives are designed as success stories, pro-jections of bounded, agentive realms of biomedicine and state control. By positioning dengue stories in their terms, the press – even in its moment of maximal state control – could be said to have snatched defeat from the jaws of victory. Dengue seemed to represent the failure of the revolution, as quint-essentially projected in medical-sanitary infrastructures and the implementa-tion of knowledge and practices inspired by public health. A theme repeated frequently in the press and by the dengue workers was that "every house and community constitutes a first step in combating the enemy" (Rassi 2002a). *Granma* reporter Rassi asserted that diverting scarce resources "would be use-less without the conscientious (*consciente*) and active participation of the popu-lation and of the social and state organizations and institutions" (Rassi 2002f).

A crucial "battle strategy" lay in "hygienic education in the family" oriented toward maintaining the cleanliness of both the home and its sur-roundings (Rassi 2002b). This arena was, of course, one of the mainstays and greatest sources of pride of the revolution (Feinsilver, 1993; Pérez, 2008; Whiteford and Branch 2009). Castro himself declared, however, that, "this battle has demonstrated that we can still greatly improve health education for the people" (*Juventud Rebelde* 2002). Both "the population" and state in-stitutions showed "social indiscipline" that "contradicted the efforts of the country's leadership, the city government, and health, community, and mass institutions" (Rassi 2002g). The focus on domestic spaces increased midway through the epidemic as officials suggested that refuse in public spaces that had provided breeding grounds for mosquitoes had largely been removed; Dr. Elia Rosa Lemus, chosen to represent the *Consejo de Estado* (Council of State) in dengue efforts, suggested that "we need to tightly contain the points where vectors concentrate, of which the greatest percentage are located in useless items in the interior of homes, like small containers, cans, and shells" (*Juventud Rebelde* 2002).

Reporters themselves embraced the work of surveillance, joining health workers in locating "undisciplined individuals and institutions who, lacking conscientiousness, violate what is indicated in our laws, threaten hygiene in the city, and play with the people's health."[5] This formulation constructed the lin-ear transmission of dengue discourse in such a way that failure to demonstrate adequate reception constituted a willful biocommunicable act of resistance. After more than four decades of public health campaigns that had been widely lauded as creating a healthy and health-consciousness national subject, Cubans were judged to be unsanitary subjects (Briggs and Mantini-Briggs 2003), making it necessary "to work day by day to *create* a hygienic culture among Cubans" (Núñez Betancourt 2002j; emphasis mine).

In the face of biocommunicable failure, two steps were necessary. One was to train "engineering troops" of laypersons, including social work students, to enter each house in La Habana and search roofs, interiors, patios, and adjacent areas, acting as "a brigade of detectives that, with the permission of the families, inspects the most hidden areas of houses and streets" (Núñez Betancourt 2002e). The Cuban health system relied not only on an immense professional work force, but on volunteers that include Committee for the Defense of the Revolution (CDR) members, students, and public workers who are ordinarily employed in other sectors.[6] These *brigadistas* searched all vessels containing water, both inside and outside houses, to see if they contained larvae or could become breeding sites – "not even little cups associated with spiritual assistance were left out" (Rassi 2002d). Inspections also involved a more general process of searching domestic spaces for signs of practices that violated "hygienic culture." Second, these "guardians of discipline" (*veladores de la disciplina*) were authorized to counter "deaf ears and laziness" with fines (Núñez Betancourt 2002f). A Rassi (2002f) article in *Granma*, whose location suggests Party backing, stated that over 17,000 fines of 100-1000 pesos were assessed from January to March.

Hygienic discipline required not just hearing these messages but listening in disciplined ways, and the press modeled the required practices of reception. The above example of the nurse's husband who achieved a perfect score on reproducing knowledge of dengue and *Aedes aegypti* modeled how the extensive body of scientific information should be assimilated. *Trabajadores* reporter Carmen Alfonso (2002) supplemented televised information on *Aedes aegypti* by publishing a quiz to test readers on how well they had learned the information. To be sure, print, radio, and television stories constantly reiterated how listening should be transformed into specific behaviors - allowing fumigators into one's home and keeping the windows closed for the required period, keeping water containers sealed, using abate when necessary, eliminating standing water in flower pots, picking up and disposing of trash, and practicing good hygiene in general. A long-standing lexicon for projecting proper modes of revolutionary reception was deployed, featuring such terms as discipline, responsibility, intelligence, seriousness, precision, honesty, quality, organization, coordination, culture, educated, and consciousness/conscientiousness (*discip-*

lina, responsabilidad, inteligencia, seriedad, exactitud, honestidad, calidad, organización, coordinación, cultura, educados, and *consciente*) – and for labeling individuals who were not listening or understanding: lack of social discipline, disorganization, deafness, laziness, errors, tolerance [of indiscipline], and lack of conscious-ness/conscientiousness (*indisciplina social, desorganización, oídos sordos, indolencia, errores, tolerancia,* and *carencia de conciencia*).

Beyond being soldiers in the dengue war, brigades modeled the trans-formation from biocommunicable failure to perfection. Núñez Betancourt (2002a) profiled brigade members Francisco Pena and José Miguel Amábiles. The article's lead depicts their initial state of ignorance: "They looked at each other and with complete humility told the authorities from the Health Minis-try: 'We don't know anything about that, but we want to do the best job possi-ble." The training received from health professionals, including videos focus-ing on dengue and *Aedes aegypti,* is credited with turning ignorant laypersons into model subjects invested with an epidemiological gaze capable of locating dengue indiscipline and reproducing biomedical discourse. Núñez Betancourt writes that the experience was transformative, becoming "an important stage in their lives" (Núñez Betancourt 2002a). Their own biocommunicable trans-formation enabled them to show other laypersons how to reproduce dengue knowledge in word and deed. There is an interesting inversion of center/ periphery, city/countryside hierarchies here in the recruitment of volunteers from other provinces to teach La Habana residents about dengue and hygiene.

Local representatives and members of mass organizations similarly as-sumed active roles in the dengue campaign. Núñez Betancourt (2002g) reports that an 81-year old woman and other residents stuck their heads of out win-dows one morning "because they wanted to thank" their local representative "for participating in this battle," thereby modeling popular responses to the communicable work performed by lay circulators of dengue knowledge. In keeping with revolutionary practice, Cuban "pioneers" (children) and adoles-cents similarly became models of dengue discourse and conduct. One child composed "a nice rap, in whose texts are messages relating to a good hygienic attitude for avoiding the dangerous mosquito" (Schlachter and García 2002). A group of young design students created animated cartoons for television as a means of "conceiving of children as protagonists in the stories told in the materials, 'because they are a principal force in every home,'" in the words of one student. "By engaging the little ones we are winning the best activists in this ongoing battle" (Núñez Betancourt 2002i).

Issues of diagnosis and treatment similarly converted families into sites of biocommunicable transmission. In another reflection of the exceptionality of dengue within Cuban bioexceptionalism, reporters told me that they generally try to avoid discussing signs and symptoms, given their goal of getting patients to visit their physicians rather than self-diagnose and self-medicate. Nevertheless, reporters quoted physicians in suggesting that "'[i]t is important that the population know the symptoms of suspected or probable cases of dengue, and especially signs of alarm'" (Rassi 2002c). Family doctors and nurses were charged with visiting dengue patients several times a day to check on their progress and to make sure that they were staying in bed and under the mosquito nets provided them. *Granma*'s de la Osa (2002) profiled dengue patient Mileidis Castillo Almeida, who modeled perfect reception of clinical discourse. Recognizing possible symptoms of dengue, which de la Osa reiterated in Castillo's voice, she immediately went to her family doctor. The neighborhood's family physicians, Carmen Córdova and Magnolia Amador, diagnosed her clinically with dengue, later confirmed by an IPK test. The physicians evaluated Castillo as "disciplined," remaining in bed with the netting in place for eight days, "conscious that in this way she would contribute to stopping the route of transmission of the disease."

The article focused on how the three protagonists played their clinical and biocommunicable roles so perfectly that Castillo wrote her physicians letters in which she thanked them not only for treating her but for constantly teaching other residents to get rid of trash, keep water tanks sealed, and help fumigators. Playing these biocommunicable roles properly strengthened family doctors' and nurses' "values of humanism, solidarity, responsibility" for their patient's health, deepened the "very personal relationship" between health professionals and residents, and produced positive feelings – even on the part of a woman experiencing dengue's very unpleasant symptoms. The accompanying photograph of the patient (holding her infant), the two physicians, and a nurse in the Castillo's living room seems to provide visual confirmation of these convergences in affective and communicable roles within an intimate space. Here is a home that has been transformed from breeding mosquitoes and viruses to reproducing laudable forms of knowledge and affect.

Complicating Biocommunicable Roles

The material that I have presented thus far suggests that biocommunicable projections of pervasive bioliteracy and biomedical citizenship are more precarious than first meets the eye. If over four decades of communicable *logros* can disappear overnight, the biocommunicable construction of ideal revolutionary citizenship seems more aspirational and precarious than stable and achieved. 2002 dengue coverage thus brings out the same sort of contradiction that appeared in my interviews with health professionals, in which they simultaneously lauded the biomedical consciousness of Cuban and accused them of being arrogant self-medicators who challenged their physicians' communicable superiority. Stories about problems encountered by health workers in confronting not just buzzing mosquitoes but Havana residents brings out an equally interesting dimension of the complexity of biocommunicable roles and how they circulate between state media and daily interactions.

In a previous article, I suggested that laypersons do occasionally emerge in Cuban health stories in roles that contradict their interpellation by doctor's orders and *logros* models (Briggs 2011). Such accounts often show laypersons exceeding the communicable spaces reserved for them by making independent clinical observations or assimilating biomedical discourse with extraordinary accuracy and completeness. Accounts in which laypersons seemed to directly challenge either biocommunicable models or the subject-positions assigned to them were particularly evident in coverage of the dengue epidemic. Castro, other politicians, and health officials declared that "every home" would constitute "a first step in this combat" (Núñez Betancourt 2002b), but many residents felt less than happy that their homes would form key sites for surveillance and forms of intervention that took material and biocommunicable forms – spraying and the "transmission" of dengue discourse, as Brotherton (2012) documents. Journalists constantly reiterated that medical specialists had remapped domestic spaces and that *brigadistas* and sprayers were supposed to bring their scientific constructions to sites of both ignorance and lack of hygiene. Nevertheless, reporters also gave voice, in multiply mediated ways, to residents who refused to accept their assigned biocommunicable role.

Castro seemed to anticipate such resistance when he exhorted the dengue control forces to be both models of revolutionary discipline and to "ensure that they are always respectful and educated with the families" (Núñez Betancourt 2002b). Expressions of resistance were mediated not only by reporters, but by the *brigadistas* and fumigators. A team comprising a physician and two nurses reported that they gave hygienic education in all houses in the neighborhood, "and even more where residents had proved themselves to be less receptive" (Núñez Betancourt 2002j). The most common modes of performing inconformity seemingly centered on fumigation, either outright by refusing to allow their house to be sprayed or opening windows prior to the prescribed 30-minute interval. An elderly founder of a local CDR reported that a youth "wanted to spray only in the living room, adducing that the mist would pass through the rest of the house" (Rassi 2002a). This youth thus resisted the technoscientific logic that guided fumigation techniques. A CDR president reported that residents who initially denied permission to sprayers "always looked for some justification" (González 2002). Their motivations included the presence of "a person who had a nervous disease," perhaps meaning that the spraying might have produced a crisis in a person with mental illness. To suggest that a mental patient might not tolerate a cloud of chemical mist points to the labor that family members performed daily, the relational and dispersed forms of care that, as Eduardo Menéndez (2005) suggests, sustain clinical medicine.

Some reportedly responded less to the materiality of spraying than to the biocommunicable model that underlay the "campaign." José Miguel Amábiles, one of the *brigadistas* I quoted above, asserted that "clearly there are always skeptics" (Núñez Betancourt 2002a). Fleeting glimpses suggested resistance on the part of the inspectors and fumigators themselves. Amábiles reported that "he was embarrassed when he arrived at each house, because he thought that he would somehow affect the comfort of residents" (Núñez Betancourt 2002a). One of the most extensive accounts of resistance emerged in Mileyda Menéndez Dávila's (2002) description of the experience of a group of social work students from Holguín in *Juventud Rebelde*. In some cases, residents refused to let the *brigadistas* enter homes or kept part of the team outside. In one case the technician entered, then a social work student, but when Ilanis Ramos tried to get through the door, "they commented that there were already too many of us, ... and that made them nervous." In what could be read as an act of either ignorance or defiance, a woman finished a soft drink while they were in the house and threw the empty can on the ground. One young *compañero* seemed to trump the other stories of communicable failure: "They threw a coconut at him from the roof." Here an official newspaper, associated with the Young Communist League, prints words and actions by Cubans who refused to be properly interpellated in their biocommunicable roles in a "campaign" that had been officially declared to be of grave concern to the Cuban state.

These articles project the interactive work required by these biocommunicable gaps. "It's not about trying to fumigate and that's all – points out José Miguel – the work has many fine points, one must be very diplomatic (*muy político*) and know how to get along with people who at a certain point don't understand and even refuse to fumigate their house." In the stories reported above, the woman picks up the can after listening to the health education information, residents finally grant access to their homes, and people are convinced that fumigation in every room is necessary. After an elderly couple initially denied access, a later attempt yielded apologies and an invitation to lunch, "When I was leaving, they gave me photos of the house and of themselves, when they were younger, so that we would never forget them" (Menéndez Dávila 2002). The hygienic micro-politics of dengue seemed to connect seamlessly with Cuban macropolitics as *brigadistas* recognized pictures of two members of the "Five Heroes" who remain "prisoners of the empire."[7] The brother of one Hero gave the students flowers, and the *brigadistas* left the other house experiencing affective proximity to "Rene" (one of the 5). Here the linear, hierarchical organization of the "campaign" seemed to get complicated as residents gave material items to the visitors. In the house of two Chinese diplomats, the wife saw that one of the young women was not feeling well. She brought the *brigadista* an herbal medicine and, while she was recovering, they exchanged knowledge of "green medicine" and beauty secrets; the resident whose domestic space was to be reinscribed by biomedical knowledge wrote with delight the visitors' names in Chinese and the words "Cuba" and "love."

Stories, materiality, and the politics of intimate spaces

What are we to make of these stories? A recent essay on dengue control brigades by Alex Nading (2012) provides insight into interactions between health workers, residents, and mosquitoes in Nicaragua. Nading points to the place of gender in these interactions, where all *brigadistas* were women and responsibility to rid households of breeding grounds fell particularly on women. This dimension is complex in the Cuban case. The *brigadistas* were both male and female, often working in mixed-gender teams; residents who attempted to block or limit access were both male and female. Generation would also seem relevant, given the youth of many of the *brigadistas* and the age of some residents who reportedly refused entrance; nevertheless, friction between younger residents and older CDR inspectors was also reported. Geography would seem also to be relevant when *bridgadistas* from the provinces presumed to impart knowledge and exercise surveillance over Havana urbanites. Race is never mentioned in the stories, but its absence might reflect representational restrictions rather than suggest its irrelevance. Accordingly, generation, gender, geography, and race might have shaped the politics of dengue-focused encounters over what are defined as domestic spaces, including ones that were not characterized as problematic, but no single binary seems to have structured them.

Nading suggests how Nicaraguan *brigadistas* embraced a variety of motives that partially displaced the bureaucratic organization of their work, hegemonic logics of public health, and discomfort surrounding the need to enter intimate spaces and conduct dengue surveillance and intervention. They enjoyed getting to know neighborhoods and conversing with a broad range of persons. Accounting for the jouissance that many experienced as *brigadistas* leads Nading to bring a missing class of actors into the analysis of these encounters who seem rather absent in the literature on Cuba – *Aedes aegypti* mosquitoes. Nading depicts *brigadistas* and residents as less engaged with a binary logic that privileges the human and relegates mosquitoes to the status of invaders from an alien realm of nature than a relational logic that draws attention to the "entanglement among people, mosquitoes, and their habitats" (2012: 74). Dengue control can thus be fun, generating pleasure through transformative recognition of the mutual becoming of people, nonhuman animals, environments, and materials that is registered through "astonishment" (Ingold 2011) produced by recognizing the presence of *Aedes aegypti* larvae in a new or hidden site. I gather from Nading's description that this fun emerged with particularly clarity in stories exchanged by *brigadistas* during lunch breaks, in which encounters with hostile canines, efforts to overcome recalcitrant residents, and new loci of mosquito reproduction were exchanged, along with objects that emerged from these interactions – from cans for recycling to mangoes proffered by more friendly householders.

Let us thus return to the accounts of resistance that emerge in the Cuban *brigadistas'* stories. Brotherton (2012) notes that many Havana residents now use part of their domestic spaces as small capitalist platforms, many of which include activities that lie outside the law. In such cases, having a *brigadista* or a member of the local CDR inspect every corner of your house or apartment might indeed produce ambivalence, to say the least. He also recounts his own resistance to the imposition of state power in the guise of a legal obligation to permit fumigation and eliminate sites of *Aedes aegypti* reproduction. This element is not missing from journalistic discourse. Núñez Betancourt reports that *bridgadistas* are aware that "if there are legal methods, some fines have been levied; the objective is to achieve the most complete hygienization possible" (2002a). When domestic spaces and their residents are not brought into conformity with Ministry maps in the course of a single "visit," resistance becomes "social indiscipline." At that point, "for those who don't want to understand or to be educated regarding the necessity of preserving the environment and collective hygiene, the application of drastic measures is the only response" (*Juventud Rebelde* 2002).

Nevertheless, Nading's accounts might open up the possibility of reading the work of Cuban *brigadistas* as similarly involving complex and shifting intersections of affects, discourses, and interests. Cuban *brigadistas* suddenly gained a new body of biomedical knowledge – which, as I have suggested, can itself produce a sense of pleasure and agency, one that is not always easily contained by health professionals. In order to gain access to intimate spaces, they often had to learn maps of domestic spaces that violated the cartographies that emerged from their training by public health professionals. Finding *Aedes aegypti* involved discovering relational cartographies in which Cuban entanglements entailed practices of consumption, water storage, gardening, health care, capitalist production, *santería*, Catholicism, and more. Rather than predetermined, simple subjects - militantly revolutionary, oppositional, or entrepreneurial - human residents who seemed to reject their revolutionary interpellation sometimes turned out to be Chinese diplomats with knowledge of green medicine or relatives of Cuban heroes.

Nading's Nicaraguan *brigadistas* seemed to have a more fixed locus for the exchange of dengue stories – a lunch break in a regular workday. Dengue in Cuba created more complex and shifting loci of narrative production and exchange. Students from the provinces shared contexts of travel, work, meals, and sleep sites to exchange narratives with one another, not to mention the "arsenal of love and principles that they brought from their lands" (Menéndez Dávila 2002). The complex networks of dengue control also involved CDR members, public health professionals, residents, and others, not to mention buses, journalists, larvae, photographs, coconuts, videos, pedagogical texts, and bureaucratic forms. In these complex, heterogeneous, and mobile contexts, it is hard to think that discourses, perspectives, interests, and motives were more homogeneous and stable than in the Nicaraguan example. Finding larvae in novel places, meeting a brother of one of the Five Heroes, receiving a gift of photographs, lunch, or herbal medicine, gaining new sources of biomedical knowledge, and, for the provincial students, the adventure of traveling to La Habana and the pleasure of teaching urbanites in the capital came together in contingent, particular ways. And all of this in an era of insecurity, the twilight zone of the Special Period, of a time of intense anxiety, a new dengue epidemic's "war" on the Cuban people, that seemed to open up the possibility of producing new subjects – human and nonhuman – and new objects, all of which were interpellated by a militaristic framing of a revolutionary campaign but none entirely predetermined by its acceptance, rejection, or self-interested manipulation. Might these be just the sort of uncertain, circumstantial, and fascinating circumstances in which new revolutionary subjects, objects, materialities, intimacies, and landscapes are being produced in contemporary Cuba?

Mediatizing the revolution

I would hope that careful readers would be voicing an objection - but these stories appear *in the media!* Crucial questions emerge: why would the state media project voices that seem to question a biocommunicable process that Fidel Castro himself inaugurated and personally supervised? Even if such stories circulated among dengue *brigadistas*, why would journalists have reproduced them in newspaper, radio, and television media? Gordy (2006) argues that even in the face of the economic exigencies and social disparities of the Special Period, many Cubans continued to reflect actively on the nature of the socialist project. Rather than a new period of pragmatism that was overdetermined by the emergence of "late Socialism" (Yurchak 2005), Gordy's work would suggest that openness, creativity, and criticism continued to be part of socialist discourse in Cuba, even if they were contained and constrained periodically in important ways, as is the case in capitalist countries today.

Remember that my journalist interviewees suggested that they were required to devote considerable energies in January-March 2002 to following the dengue story. "Popular participation" was not a simple process during the dengue epidemic. Indeed, the team from the leading Cuban biomedical institution IPK, which probably had the deepest engagement with such efforts, reported that even highly focused efforts to foster popular participation tended to reinforce the authority of CDR leaders rather than disperse power (see Sánchez et al. 2008). But calls for popular participation did provide journalists with a motive for spending time with *brigadistas*, CDR members, and residents in their search for dengue stories. Networks of *brigadistas*, mosquitoes, residents, lids for water vessels, and old bottle caps thus came to include journalists, who became audiences for the stories exchanged by persons involved in dengue control efforts.

Their presence should not naively be presented, however, as an explanation as to why journalists would insert *brigadistas* stories in *Juventud Rebelde, Trabajadores*, and *Granma*. We should thus reflect, I think, about why what might seem to be tales of resistance to Fidel Castro's own announced biocommunicable program and its human and material presence might appear in these highly visible media. Here we have, I suggest, a complex interaction between scales and sites in the emergence of complex, heterogeneous subjects who embody a key revolutionary project – defending *la patria* in a war on dengue – but who do so through a variety of motives, positionalities, materialities, and interests. There is room here for journalism, for reporters looking for good stories, which is how the media professionals presented this work to me. At the same time, during the time of particular insecurity during the uncertain late Special Period era, the state media used its pages and broadcasts to feature stories that seemed to open up spaces for the emergence of complex subjects and objects who entered into revolutionary projects without being predetermined by them, some of whom seemed to resist their revolutionary interpellation, and some of whom were nonhuman actors who buzzed (or swam) at the margins of official discourses and materialities.

Conclusion: Contingent Narratives of Contemporary Cuba

It would be easy to suggest that this critical opening was too short to warrant serious consideration. Indeed, Castro returned to the Karl Marx on 27 March 2002, and his triumphal words were relayed by the two *Granma* reporters who had taken particularly active roles in reporting on dengue, Alberto Núñez Betancourt and Reynold Rassi (2002), as well as journalists for the main television channel and two radio stations. Castro declared that no new cases were being reported and that the *Aedes aegypti* population had been reduced to a minimum.[8] This declaration enabled dengue coverage to move back to *logros* format. First Party Secretary in La Habana Esteban Lazo produced statistics that detailed the different groups that had participated and listed their numbers, before replying to Castro: "As you said, Comandante, the task was difficult and complex. But here we are today with the mission complete and having demonstrated one more time that the right to life and to health are completely guaranteed in our homeland (*Patria*)." The importance of the achievement was tied to the "scientific level" of the strategy on which it was based. Castro then declared the political effects of the campaign: "The people demonstrated again that they are capable of doing impressive things, of fighting on various fronts at the same time and winning" (Núñez Betancourt and Rassi 2002). As direct state control over health news faded away along with the ubiquitous logo of *Aedes aegypti* in the crosshairs, anxious questioning of state and popular biocommunicable compliance faded and the teleological and triumphalist tones of doctor's orders and *logros* rhetorics reemerged.

Indeed, a national biocommunicable success provided another opening for Cuban medical diplomacy: international health organizations could extract lessons from the campaign launched by Cubans. Castro projected that this circuit was already circulating Cuban dengue discourse internationally. Reporters turned to international officials to validate the campaign – and thus Cuban bioexceptionalism. Reporter Flor de Paz (2002) interviewed the Pan American Health Organization's country representative Patricio Yépez. At the same time that he projected dengue as "the principal health problem in the Americas," thereby placing Cuba within a widespread problem, Yépez buttressed Cuban bioexceptionalism: "It is necessary to link political will and decision-making, inter-agency cooperation, popular participation, and the availability of resources with which to combat the problem massively, and not all of these elements come together elsewhere like they do in Cuba" (Paz 2002). Yépez concluded that Cuba's methodology "will constitute a convenient point of reference for other countries to take into account." Indeed, an article in *The Lancet* echoed the language of bioexceptionalism in citing dengue control as pointing to why many "western scientists admire, with some reservations, the priorities and achievements of Cubans in tackling health problems that similar countries find intractable" (Sansome 2007). Although journalists, as I noted, claimed that government officials require them to keep dengue in the news, coverage dropped rapidly after the epidemic had officially ended; stories tapered off in April-July 2002, and by year's end very few articles or television broadcasts were appearing.

We might thus be tempted to sweep dengue coverage under the historical rug. Although I would resist this tendency, neither I do not want to suggest that these stories provide us with some sort of paradigm for projecting a new consistent alignment of materialities, subjects, and political-economic arrangements in Cuba. I like the convergence between storytelling practices among particular CDR members, *brigadistas*, health professionals, and Havana residents and narratives that emerge in the official press precisely because of their contingent and unstable character. Dengue opened up cracks in pedagogical and *logros* models of biocommunicability, both through official admissions of their aspirational and unstable character and in the space devoted to efforts to fashion new subjects who operate within but are not totally configured by the subject-positions they project. As I argued above, these fissures did not emerge *de novo* in January 2002 but form part and parcel of Cuban biopolitics. They are not simply popular or official but rather challenge any such binary, not through socialist suppression or nascent capitalist self-interest but rather through complex, shifting intersections across locations and scales.

I have tried to open up a space here for creating narratives of the emergence of medical subjects and objects and Cuba that resist synecdochic logics and totalizing narratives but follow contingents, complexities, and heterogeneous perspectives and interests. In doing so, I have tried to point out how medical anthropologists might engage journalistic mediation not as an ethnographic supplement or as a unified and all-powerful field of power. My rather peculiar object, media coverage of the 2002 dengue epidemic, did not reveal the Party's efforts to either suppress or amplify preexisting, officially constituted biomedical subjects and objects. Nor did it reveal hidden fields of popular efforts to resist official truths. Mosquitoes, *brigadistas*, coconuts, abate, and herbal medicines would rather seem to provide us with more modest and limited views, rather like the sudden discovery of hundreds of *Aedes aegypti* larvae in a discarded bottle cap, of how the emergence of new Cuban subjectivities, materialities, and politics is, to use Ralph Ellison's (1972: 152) powerful phrase, "hidden right out in the open."

References

Alfonso, C.R. (2002) 'Táctica y estrategia para vencer al enemigo', *Trabajadores*, 14 January 2002, 8.

Anderson, W. (2006) *Colonial Pathologies: American Tropical Medicine, Race, and Hygiene in the Philippines*, Durham: Duke University Press.

Arnold, D. (1993) *Colonizing the Body: State Medicine and Epidemic Disease in Nineteenth-Century India*, Berkeley: University of California Press.

Briggs, C.L. (2011) '"All Cubans are doctors!": news coverage of health and bioexceptionalism in Cuba', *Social Science and Medicine*, 73: 1037-1044.

Briggs, C.L. and Hallin, D.C. (2007) 'Biocommunicability: the neoliberal subject and its contradictions in news coverage of health issues', *Social Text*, 25: 43-66.

— (2010) 'Health reporting as political reporting: biocommunicability and the public sphere', *Journalism: Theory, Practice, and Criticism*, 11:149-165.

Briggs, C.L. and Mantini-Briggs, C. (2003) *Stories in the Time of Cholera: Racial Profiling in a Medical Nightmare*, Berkeley: University of California Press.

Castro, F. (2005) 'Ustedes honran la noble profesión médica', *Trabajadores*, 5 September 2005, 9.

Clarke, A.E., Shim, J.K., Mamo, L., Fosket, J.R. and Fishman, J.R. (2003) 'Biomedicalization: technoscientific transformations of health, illness, and U.S. biomedicine', *American Sociological Review,* 68: 161-94.

De la Osa, J. (2002) 'Historia para contar', *Granma,* 18 February 2002, 6.

Díaz Briquets, S. (1983) *The Health Revolution in Cuba,* Austin: University of Texas Press.

Eckstein, S.E. (1994) *Back from the Future: Cuba under Castro,* Princeton: Princeton University Press.

Ellison, R. (1972) *Invisible Man,* New York: Vintage.

Feinsilver, J.M. (1993) *Healing the Masses: Cuban Health Politics at Home and Abroad,* Berkeley: University of California Press.

Foucault, M. (1976[1978]) *The History of Sexuality.* Vol. 1: *An Introduction,* Robert Hurley (trans.), New York: Vintage.

— (1997) *Ethics: Subjectivity and Truth,* Paul Rabinow (ed.), Robert Hurley et al. (trans.), New York: New Press.

González, A.M. (2002) 'La comunidad acecha a un enemigo común', *Trabajadores,* 21 January 2002, 9.

González Rubio, D., Castro Peraza, O., Rodríguez Delgado, F., Portela Ramírez, D., Garcés Martínez, M., Martínez Rodríguez, A., Rodríguez Bada, N. and Guzmán Tiradoet, M. (2008) 'Descripción de la fiebre hemorrágica del dengue, serotipo 3, Ciudad de la Habana, 2001-2002', *Revista Cubana de Medicina Tropical,* 60: 48-54.

Gordy, K. (2006) 'Sales + economy + efficiency = revolution: consumer capitalism, and popular responses in Special Period Cuba', *Public Culture, 18*: 383-412.

Gubler, D.J., and Kuno, G. (eds.) (1997) *Dengue and Dengue Hemorrhagic Fever,* Wallingord, UK: CAB International.

Guzmán, M.G., Peláez, O., Kourí, G., Quintana, I., Vázquez, S., Pentón, M., Avila, L.C. and el Grupo Multidisciplinario para el Control de la Epidemia de Dengue 2001-2002 (2006) 'Caracterización final y lecciones de la epidemia de dengue 3 en Cuba, 2001-2002', *Revista Panamericana de Salud Pública/Pan American Journal of Public Health,* 19: 282-289.

Guzman M.G., and Kourí, G. (2002) 'Dengue: an update', *Lancet Infectious Diseases,* 2: 33–42.

Hernández-Reguant, A. (ed.) (2008) *Cuba in the Special Period: Culture and Ideology in the 1990s,* New York: Palgrave.

Hirschfeld, K. (2007) *Health, Politics, and Revolution in Cuba since 1898,* New Brunswick, NJ: Transaction.

Ingold, T. (2011) *Being Alive: Essays on Movement, Knowledge, and Description*, New York: Routledge

Juventud Rebelde. (2002) 'Educar para proteger al pueblo', *Juventud Rebelde* 22 February 2002, 1.

Kourí, G., Guzmán, M.G. and Bravo, J. (1986) 'Hemorrhagic dengue in Cuba: history of an epidemic', *PAHO Bulletin*, 20: 24-30.

Kourí, G., Guzmán, M.G., Valdés, L., Carbonel, I., Rosario, D. del, Vazquez, S., Laferté, J., Delgado, J. and Cabrera, M.V. (1998) 'Reemergence of dengue in Cuba: a 1997 epidemic in Santiago de Cuba,' *Emerging Infectious Diseases*, 4: 89-92.

Khun, S. and Manderson, L.H. (2007) 'Abate distribution and dengue control in rural Cambodia', *Acta Tropica*, 101: 139–146

Latour, B. (1987) *Science in Action: How to Follow Scientists and Engineers through Society*, Cambridge, MA: Harvard University Press.

Marquetti, M.C., Suárez, S., Bisset, J. and Leyva, M. (2005) 'Reporte de hábitats utilizados por Aedes aegypti en Ciudad de La Habana, Cuba', *Revista Cubana Medicina Tropical*, 57: 159-161.

Menéndez, E.L. (2005) 'Intencionalidad, experiencia y función: la articulación de los saberes médicos', *Revista de Antropología Social*, 14: 33-69.

Menéndez, E.L., & Di Pardo, R.B. (2010) *Miedos, riesgos e inseguridades: el papel de los medios, de los profesionales y de los intelectuales en la construcción de la salud como catástrofe*, Mexico: CIESAS.

Menéndez Dávila, M. (2002) 'Vivencias para toda la vida', *Juventud Rebelde*, 20 February 2002, 8.

Mitchell, T. (2002) *Rule of Experts: Egypt, Techno-Politics, Modernity*, Berkeley: University of California Press.

Nading, A.M. (2012) 'Dengue mosquitoes are single mothers: biopolitics meets ecological aesthetics in Nicaraguan community health work', *Cultural Anthropology*, 27: 572-596.

Nichter, M. (2008) *Global Health: Why Cultural Perceptions, Social Representations, and Biopolitics Matter*, Tucson: University of Arizona Press.

Núñez Betancourt, A. (2002a) 'Jaque constante al Aedes aegypti', Granma, 9 January 2002.

— (2002b) 'Ofensiva contra el Aedes aegypti', *Granma*, 12 January 2002.

— (2002c) 'No trasladar los focos a las calles', *Granma*, 17 January 2002.

— (2002d) 'En busca de un trabajo integral', *Granma*, 19 January 2002.

— (2002e) 'La calidad bajo control', *Granma*, 1 February 2002.

— (2002f) 'Veladores de la disciplina', *Granma* 9 February 2002.

— (2002g) 'Líderes con el concurso de todos', *Granma*, 13 February 2002.

— (2002h) 'Participó el Comandante en Jefe en el chequeo diario de la ofensiva antivectorial', *Granma*, 22 February 2002.

— (2002i) 'Creación de los comunicadores visuales', *Granma*, 23 March 2002.

— (2002j) 'El camino de una cultural sanitaria', *Granma*, 17 January 2002.

Núñez, A., and Rassi, R. (2002) 'El dengue ha sido erradicado de Cuba', *Granma*, 28 March 2002.

Paz, F. de (2002) 'Una referencia para otros países', *Juventud Rebelde*.

Pérez, C. (2008) *Caring for Them from Birth to Death: The Practice of Community-Based Cuban Medicine*, Lanham, MD: Lexington.

Pérez-Stable, M. (1999) *The Cuban Revolution: Origins, Course, and Legacy*, New York: Oxford University Press.

Rassi, R. (2002a) 'La familia y la comunidad: escalón primario en el combate contra el Aedes', *Granma*, 23 January 2002.

— (2002b) 'Indisciplinas sociales en la higiene', *Granma*, 23 January 2002.

— (2002c) 'Un ejemplo en la atención al ingreso domiciliario', *Granma*, 11 February 2002.

— (2002d) 'Una importante tropa ingeniera', *Granma*, 15 February 2002.

— (2002e) 'La eliminación de los criaderos en el mismo centro del problema', *Granma*, 19 February 2002.

— (2002f) 'Educación sanitaria: la batalla permanente', *Granma*, 13 March 2002.

— (2002g) 'Indisciplinas sociales en la higiene', *Granma*, 18 March 2002.

Rigau-Pérez, J.G., Clark, G.G., Gubler, D.J., Reiter, P., Sanders, E.J. and Vomdam A.V. (1998) 'Dengue and dengue haemorrhagic fever', *The Lancet*, 352: 971-77.

Rivero, Y.M. (2012) 'Watching TV in Havana: revisiting the local/global television past through the lens of the television present', in *New Agendas for Global Communication*, (eds.), London: Routledge.

Sánchez, L., Pérez, D., Alfonso, L., Castro, M., Sánchez, L.M., Van der Stuyft, P. and Kourí, G. (2008) 'Estrategia de educación popular para promover la participación comunitaria en la prevención del dengue en Cuba', *Revista Panamericana de Salud Publica*, 24: 61–9.

Sansome, C. (2007) 'Tackling infectious disease in Cuba', *The Lancet Infectious Diseases*, 7: 376.

Schlachter, A., and García, P.A. (2002) 'No habrá descanso en esta batalla por la salud del pueblo', *Granma*, 26 February 2002.

Whiteford, L.M. and Hill, B. (2005) 'The political economy of dengue in Cuba and the Dominican Republic', in G. Guest (ed.) *Globalization, Health, and the Environment*, Walnut Creek, CA: Altamira Press.

Whiteford, L.M. and Branch, L.G. (2009) *Primary Health Care in Cuba: The Other Revolution*, Lanham, MD: Rowman & Littlefield.

Yurchak, A. (2005) *Everything was Forever, until It Was No More: The Last Soviet Generation*, Princeton: Princeton University Press.

Endnotes

1   Staff members of the Instituto de Medicina Tropical "Pedro Kourí," Radio Progreso, the Centro Nacional de Promoción y Educación para la Salud, Cubavisión, the Escuela Nacional de Salud Pública, Granma, Juventud Rebelde, and Trabajadores, the Universidad de La Habana, various clinics and polyclinics, and other institutions provided crucial assistance, along with Cubans who welcomed me into their living rooms. I thank my collaborators in the media and health project, particularly Jaime Breilh, Daniel Hallin, Clara Mantini-Briggs, Eduardo Menéndez, and Hugo Spinelli, for conversations that illuminated key issues. Karen Hernández, Monica Moncada, and Kayle Rodríguez helped transcribe interviews and analyze the media collection. Financial assistance was provided by the University of California, Berkeley, the University of California, San Diego, and the Salus Mundi Foundation. Nancy Burke, Mark Nichter, and three anonymous reviewers provided valuable comments.

2   Given the focus on 2002, "Castro" refers to Fidel Castro, not Raúl.

3   I also sampled coverage 2005-2008, including these newspapers and health-related broadcasts on Radio Progreso and television station Cubavisión (154 additional stories). I consulted media archives from the 1960s through the present and regional newspapers. Although this material informs the broader analysis, it is not relevant to the 2002 dengue coverage.

4   All interviewees were given pseudonyms. The translations are my own, although I consulted bilingual medical professionals in translating technical terms.

5   It is interesting to note the similarity here to the roles assumed by reporters for commercial press outlets in pre-revolutionary Venezuela during the 1991-1993 cholera epidemic, who seemed to extend the work of epidemiological surveillance into homes, shantytowns, sites where food, drink, and other goods were sold on streets, and indigenous communities (see Briggs and Mantini-Briggs 2003).

6   See Feinsilver (1993) and Brotherton (2012) on *brigadistas* and their importance to public health in Cuba.
7   The Five Heroes are Cubans arrested while investigating U.S. residents involved in terrorist acts against Cuba. They remain imprisoned in the United States and have become a cause célèbre on the island.
8   See Guzmán et al. (2006) and González et al. (2008) on the 2001-2002 epidemic.

2

Disconnecting the Mind and Essentialized Fare: Identity, Consumption, and Mental Distress in Santiago de Cuba[1]

Hanna Garth

While telling me parts of his life story, Patricio,[2] an 86-year-old, middle-class, white man who had lived in Santiago de Cuba all of his life, became visibly distressed as he began to talk about food in Cuba today. His brow furrowed, his voice became softer and deeper, and he finished his sentences with deep lamenting sighs. Patricio had just told me of his experiences as head chef in a fancy hotel in the 1950s and the conversation shifted to his reflections on a recent dinner:

> My daughter had come over from France, she lives there now and only visits once a year. This was our special night out; we went to one of the nicer *paladares* [restaurants] in [Santiago]. I ordered the fish, because we hadn't had fish in the rations for a while. When the plate came I was so disgusted, it was just a fish on a plate, no spices, no garnish, no presentation, no art. We paid so much money for that dinner and it was worse than what I would have made at home. It was a disgrace, an insult. My daughter, who knows well the fancy cuisine of Europe, will never know the Cuban food that I grew up with, that I made in the hotels, she will only know that overpriced, overcooked, ugly fish.

Patricio's disappointment with changes in food preparation and presentation reveal an important sense of nostalgia and loss of grace common among my research participants. His comments are also illustrative of a more common frustration with the difficulty of accessing quality food and other goods that

Santiagueros view as integral to attaining a decent quality of life in contemporary Cuba (Garth 2009; Hernandez 2008; Wilson 2009). Patricio's shame and disappointment with this meal are augmented due to the fact that he, like many other Santiagueros, connects pride in his national cuisine with his sense of Cuban identity. A demonstration of David Sutton's (2001) notion that if "we are what we eat," then "we are what we ate," Patricio reveals the problem examined in this article: how does the link between food and identity change as the foods people have historically linked to their identity become more difficult to access?

Integrating psychological anthropology with studies of consumption, I illustrate the discursive and affective ways present-day Cubans negotiate the tensions between idealized notions of food and identity and state-determined provisioning. I use the term *identity* to describe how Cubans living on the island today understand who they are as Cubans through dynamic and ever-changing interpretations and evocations of their national history in relation to their contemporary lives (Hall 1996; Khan 2004; Wilk 1999). I examine the self-conscious struggle to maintain what I refer to as "essentialized fare" in the face of the scarcity of particular foods. "Essentialized fare" is my interpretation of the local notion that Cubans were made to eat particular foods that are often thought of as fundamentally Cuban in origin. This includes idealized notions that Cuban bodies are best suited to digest particular foods, that Cuban taste is best suited to certain food items, and that certain foods are an integral part of Cuban national and cultural identity. Santiagueros often eat other kinds of foods in practice, but they discursively maintain that Cubans *should* eat a certain way. I contextualize the ways in which Santiagueros link ideas about "decent" food to their sense of Cuban identity – as did Patricio in regretting his expatriate daughter's experience with their unfulfilling meal – within a broader social struggle to maintain cultural identity (de Certeau 1984). The sense of a decent meal is integral to participants' understandings of "what really matters" in maintaining a sufficient quality of life and what is at stake in leading an adequate life (Kleinman 2006). I subsequently analyze the role of the practice of "disconnecting the mind" (*desconectar la mente*), a method used to cope when the struggle to acquire these goods becomes too difficult. "Disconnecting the mind" or simply "disconnecting" involves a mental process of not thinking about difficult or stressful situations. Rather than reflecting upon struggles or problems, the practice of "disconnecting the mind" involves removing stressful thoughts and situations from one's immediate attention and in turn focusing on positive things and issues over which one has more control. I argue that Cubans cope with this food-related distress over their national identity through a specific mechanism they call "disconnecting the mind," thus revealing how the coping mechanism of "disconnecting the mind" illuminates the importance of the identity – consumption – mental distress nexus.

Food, Identity, and Mental Distress

A large body of research has shown that food consumption, in particular, is conditioned by a wide range of symbolic meanings, from class distinction to religious practice to cultural preferences and nationalism (Allen 2002; Appadurai 1988; Askegaard 1999; Barthes 1997; Bourdieu 1984, 1993; Gofton 1986; Mintz 1996; Premat 1998; Sahlins 1990). More specifically, a number of scholars have described the ways in which consumers align themselves with particular products and food choices as a way of making claims about their identity (Bentley 2001; Berdahl 2005; Bott 1987; Garth 2013; Kahn 1986; Miller 1995; Scholliers 2001; Sutton 2001; Weismantel 1989; Wilk 2006; Whitehead 2000). For instance, Roseberry (1996) has demonstrated a link between class identity and the growth of a specialized coffee market in the United States. Caldwell's (2002) work in post-Soviet Russia shows how consumers orient toward the nationalist values and political ideals associated with specific food choices in Moscow. Counihan and Kaplan (2004) connect certain foods and their manner of consumption with self-ascribed gender identity. Building upon this work, I show how the social and cultural meanings that people inscribe onto Cuban cuisine today are integrated into Cuban identity through both individual and social practices of food acquisition and consumption.

In particular, Caribbean scholars and food scholars alike have connected Caribbean foodways – the cultural, social, and economic practices relating to food production and consumption – with efforts to define and mobilize national identity (Derby 1998; Wilk 2006). Food and agriculture have long been central to Cuban national and local identity (Dawdy 2002). For many Cubans, it is not just actual foods that are tied to Cuban identity, but also the *struggle* to acquire food and other goods, referred to as "*la lucha*." The term *la lucha* originally dates back to Cuba's struggle for independence from Spain in the 1800s.[3] Since Cuban independence from Spain, the concept of *lucha* has been tied to *Cubanidad* (Cuban-ness) and *patria* (literally "father-land"). These ideas have been continually infused into Cuban national identity, both before and after the 1959 revolution. As Louis A. Pérez Jr. has recently written, the logic of *patria* in Cuba today extends from a "19th-century conception of nationhood as a 'place-bound source of self-identification'" (2009: 12). The pursuit of *patria* and anti-imperialism are exemplified through the ubiquitous Cuban refrain: "to die for the *patria* is to live" (see for example Gordy 2006). The process of acquiring the types of foods that Cubans view as absolutely essential to Cuban cuisine and identity is also often described as a *lucha;* the struggle itself is seen as a central part of the pursuit of *patria* (Pérez 2009; Brotherton 2005). In post-Soviet socialist Cuba, the term *la lucha* describes both ongoing daily efforts to achieve the goals of revolutionary society and efforts that individuals must go through to get beyond the controls and barriers that have been in place since the revolution. That is, to *resolver* (to resolve) their everyday problems.

Socialist ideals articulate particular ways of disciplining and nourishing citizens through certain forms of subjectivity (Anagnost 1995; Gal and Kligman 2000; Kligman 1998). Consumption here is tied not only to a broader sense of Cuban identity, but also to the ways in which consumption and identity are shaped by the socialist state. The rationing system provides an excellent example of how state attempts to meet the population's needs within conditions of economic scarcity shape Cuban consumer identity through providing foods that are easy for the state to acquire and distribute, which at the same time are items that the population is willing and able to consume. More than a response to scarcity, the rationing system was established and designed to enhance equality in Cuba (Benjamin et al. 1984). Both practically and ideologically, food has occupied a special position within the experiences of Cuban state socialism, with respect to both citizens' relations with the state and relations between citizens (see Caldwell 2004). On the other hand, food scarcity in everyday life frequently prompts household struggles to find food, as during the Special Period of economic hardship in the 1990s (Holgado Fernández 2000). These struggles may ironically subvert the socialist connection between food and identity, as in class-tinted efforts by Santiago women to find the foods and goods that they feel are necessary to maintain a decent standard of living (see Pertierra 2007, 2008).

Such tactical struggles on the part of individuals, together with the fact that scarcity in Cuba usually is of particular food items rather than a general lack of food, point to a rupture between constrained food consumption practices and local concepts of identity and values (see Wilson 2009, 2010). Wilson (2009, 2010) describes Cubans' use of irony and humor to ameliorate such ruptures. When situated within the larger context of socialist ideals and post-

Soviet privations, such reparative activities indicate that food scarcity has psychosocial repercussions for Cuban identity (see for example Tanuma 2007).

In fact, the struggle to maintain cultural identity and traditions has been linked more generally with mental distress (Adler 1995; Desjarlais et al. 1995; Hadjiyanni and Helle 2009). At least as it is used in the context of food acquisition, *lucha* is a Cuban "idiom of distress[4]" (Nichter 2010). Such idioms:

> are socially and culturally resonant means of experiencing and expressing distress in local worlds. They are evocative and index past traumatic memories as well as present stressors, such as anger, powerlessness, social marginalization and insecurity, and possible future sources of anxiety, loss and angst. Idioms of distress communicate experiential states. . .from the mildly stressful to depths of suffering that render individuals and groups incapable of functioning as productive members of society. In some cases, idioms of distress are culturally and interpersonally effective ways of expressing and coping with distress . . . . (Nichter 2010: 405)[5]

For a cultural symbol like food to become a socially significant idiom of distress that indexes subjective experience, an individual must incorporate the symbol into their everyday emotional concerns (Obeyesekere 1981; Throop 2003). When it is, changes in the material resources and personal symbols previously used in the construction of identity may become a source of distress. This can be so even if those changes are objectively improvements (Desjarlais et al. 1995). Unlike settings in which *hunger* becomes a cultural idiom (see for example Davis and Tarasuk 1992; Hastrup 1993; Scheper-Hughes 1993; Pottier 1999), among my research participants hunger itself is not central the source of mental distress, but rather the scarcity of particular food items defines an important idiom of distress in Cuban identity today. This is not to say that hunger is not a real risk in Santiago and other parts of Cuba, especially if a combination of food acquisition tactics is not skillfully pursued. In the Cuban case, despite the increasing availability of basic food staples in the past decade, for many Santiagueros food is still the object of emotive longings and desires for consumer goods and remembered lifestyles of the past. In recent years, particularly in Santiago, food scarcity and desires for particular food items linked to local identity has been a major source of stress and struggle that often leads to the need to "disconnect the mind," a local practice of alleviating mental distress.

Research Site and Methods

This research took place in the southeastern part of Cuba in the city of Santiago de Cuba, an ethnographically rich urban setting in which to study the experience of contemporary Cuban life through the particular lens of food acquisition. Santiago is part of the *Oriente* region of Cuba. Many people from *Oriente* self-identify as *guajiros,* or peasants. Locals conceptualize Santiago as an urban center with a rural pace of living. Santiago is known for its vibrant "Afro-Cuban"[6] culture. Many Santiagueros are proud to continue what they perceive to be African traditions through music, dance, and ritual forms, and through culinary practices as well. Santiagueros connect the foods they eat today with their understandings of the foods their ancestors ate in the past; they derive an important sense of identity from contemporary connections to remembered pasts.[7] Santiago provides an urban setting through which to view food acquisition and symbolism in a very distinct cultural and social context compared to Havana, which has attracted most ethnographic attention and is an object of comparison for Santiagueros themselves. Indeed, the techniques employed to find, process, prepare, serve, and consume foods are geographically variable, with distinct histories and meanings in different research settings.

The data for this study were collected over seven non-consecutive months in 2008, 2009, and 2010. With the official sponsorship and generous help of the *Casa del Caribe,* I was able to access families to participate in this household-based ethnographic study. Using participant observation and interviews, I studied the processes involved in and discourse surrounding the daily acquisition and consumption of food for Santiago households in neighborhoods across the city. Through these methods I observed what foods were served, how they were prepared, how people spoke about and reflected upon these foods, and their access to them as their daily meals were being consumed. While food was the central focus of this study, I also developed a thorough understanding of many other aspects of everyday life in Santiago de Cuba, including housing issues, child rearing, finances, work, family relations, etc.

I conducted informal, semi-structured, audio-recorded interviews with a diverse cross section of 25 adult residents of Santiago. Participants were identified through word of mouth and networking with my contacts in the city. I conducted all of the research in Spanish and transcribed the interviews myself. The research participants had a wide age range, from 25 years old to 75 years old, which gave participants varying levels of familiarity with the historical changes in the Cuban food system. Just over half of the participants were sixty years old or older and were able to recollect what life was like in Santiago before and after the 1959 revolution. All of the participants have lived in the province of Santiago de Cuba throughout their lives, and eighty percent of them have always resided within the city of Santiago. Nevertheless, two-thirds retain close ties to family members who live on farms in the nearby country-side. In addition to ascertaining participants' concerns about food availability, I gained a deeper understanding of local food folklore and symbolic interpretations of the local cuisine.

The Struggle for "Essentialized Fare"

Expressions of a remembered traditional cuisine

In Santiago I found that it was common to hear sweeping generalizations about Cuban taste, including statements such as *"al Cubano le gusta el puerco"* (the Cuban likes pork) or comments about how Cubans could never be vegetarian, *"no somos así"* (we aren't that way).[8] As Omaro, a 58-year-old black middle class Cuban, put it:

> Well if you are studying food in Cuba – Cuban food that is – all you need to know about is beans and rice. Well. . . we have that joke about the three failures of the revolution - breakfast, lunch, and dinner[9]. . .[laughter]. . .but seriously what you need to know about is rice and beans - what we call here *Congrís* - and then pork and *viandas* that's all we eat because that is *our* food. What the Cuban people live, our way of living, is *congrís,* pork, and *viandas.*

Omaro's reference to the popular complaint of the 1959 revolution's failures to provide the types of foods that Cubans desired illustrates how important many feel it is to maintain this particular Cuban diet.

Similar to Omaro's comments, María Isabel, a 67-year-old, mulatta retired doctor who lives in a middle class Santiago neighborhood expressed to me her own view of the Santiaguero struggle to maintain local food traditions:

> And for example, we have another dish that we have tried so hard to pre-serve it as it is - the *hallaca* - in the west they call it *tamal,* but we have to maintain our *hallaca.* So we have entered into a state of a lot of energy and effort around this, we keep trying to find a way to overcome these difficul-

ties, to continue living our way of life. But without forgetting that we have a very large psychological blow (*golpe psicológico*) in many things.

For María Isabel, the struggle to maintain this particular local way of eating is so deeply tied to her identity that she casts the difficulties of maintaining this diet as a *"golpe psicológico"* (psychological blow or attack) (see below). In the first part of María Isabel's statement we see that she is quick to note the importance of preserving a distinct Santiago dish, which she elaborates later in the interview as much more than a linguistic distinction between *hallaca* and *tamal*. She explained that:

> What is most important is that our food is grown here on the *Oriente* soil, which is so fertile with a history - with our history - of the blood and the *lucha* of our African, French, Haitian, and Cuban ancestors and years of *lucha* and victory that give the corn grown in the region with a sweeter flavor and juicier kernels. And also the people who sell corn here in Santiago they still grind the corn by hand for *hallaca,* and this gives it the flavor that makes it special to us.

While her sentiments reveal a highly romanticized view of the relationship between Cuban history and contemporary Cuban foods, María Isabel communicates here how much she values particular aspects of the food production and preparation for their connections to her Santiaguera identity. She clearly prefers foods grown not only on Cuban land, but on *Oriente* land as not only better tasting but as morally laden with Santiaguero work ethics and values. She casts the preservation of regional dishes as a type of struggle that she continues to put energy into despite difficulties. María Isabel's focus on a dish that is not an everyday food and does not use the typical staple ingredients of Cuban cuisine was also quite common. I found that although participants did complain about the scarcity of staples such as rice or beans, in practice the foods that they really struggled to acquire were not staples but rather ingredients that were important for flavoring such as spices, oil, or ingredients for less common side dishes. These types of foods are part of essentialized fare as well.

Although the romanticization of traditional food consumption habits is quite common, some participants note that food habits have changed over time (for further analysis of ideological responses to food consumption, see Caldwell 2002; Klumbyté 2010.) For example, towards the end of one of my interviews with Martin, a young, working class, black Santiaguero who raises pigs on his roof, I asked him what was his favorite meal. He smiled and replied that his preferred meal was: "*Congrís* [white rice and black beans], *yuca, tostones* [deep fried mashed green plantains], and pork with seasonal vegetables, of course." When I naively responded with the statement, "like every Santiaguero," Martin quickly corrected my assumption. He pointed out that now, with the influence of tourists, television, and fancy Cuban convertible peso (CUC) restaurants throughout Cuba, many Santiagueros want to "change what they eat," and that although many cannot afford to eat in the tourist restaurants they try to replicate this foreign way of eating in their homes. He used the example of the popularity of Chinese food in Cuba, the fact that many families now eat pizza and spaghetti at home, and what he observes as an increasing desire for the Euro-American aesthetic, not only with regard to food, but in dress, music choice, and "attitude".[10] He pointed out that although people eat these types of foods:

> These foods are not our way of eating. The Cuban body wants our traditional foods that is what we have to eat. People eating these foreign foods and chemical foods, that is why there is so much sickness, so much cancer. We have to eat our own foods to maintain our health.

In response to a question regarding how she felt about the Cuban government adding soy to the ground beef in the rations, Carolina a 35-year-old, working class, white Santiaguera who works in the local soy processing plant explained that:

> Soy is very aggressive for the Cuban body, for the stomach, it's very hot and since the climate here is so hot our bodies are not adjusted to such hot foods. Not all bodies are the same no? Some people, like from northern climates can eat soy, but not Cubans.

Carolina's classification of soy as a food that is too hot for Cuban bodies, a humoral pathology reference, again asserts the idea that Cuban bodies were made to consume certain "Cuban foods," a critical underlying concept for the local understanding of essentialized fare. Both Martin and Carolina allude to the notion that the Cuban body is best suited to certain kinds of foods. Martin's comments about what he perceives as an increase in the consumption of "foreign" foods may illustrate a discrepancy between how Santiagueros actually eat in practice versus the idealized, essentialized cuisine that they claim to desire. Although foods such as Chinese food,[11] pizza and spaghetti may, in some situations, be cheaper and more easily accessed than traditional Cuban cuisine, my research reveals that the consumption of these foods may actually be a source of ambivalence and shame in light of local ideologies that push the consumption of *Cuban* food.[12] The consumption of "foreign" foods thus negates those ideologies.

Food acquisition and the food rationing system in Santiago de Cuba

In an interview with 54-year-old Gloria, a white middle class Santiaguera, she shared with me comments that I heard repeated in nearly every interview I conducted:

> The ration is absolutely fundamental for many people to survive, but it doesn't last. It's just a few days of rice and beans and a little bit of meat. It's not enough, but it's fundamental for life here.

Over the past three years conducting research in Santiago, I have consistently found that the foremost concern of many Santiagueros is the supplementation of their monthly food ration with food items that they view as essential to maintain their ideal of Cuban cuisine. The concern over food acquisition that I observed was often articulated as a *lucha*, which, as I will show later, is as much about a desire to maintain what Santiagueros perceive as Cuban and *Oriente* culinary traditions deeply linked to local identity as it is about attaining adequate nutrition. Research participants expressed a particular desire for the following foods: pork, beef, chicken, additional rice, black beans, red beans, chickpeas, eggs, milk, and, probably most importantly, cooking oil.[13] Several people, many of whom were over sixty years old, lamented the loss of herbs and spices used in recipes before the revolution, specifically that access to items used to flavor foods (such as cumin, oregano, culantro,[14] onions, green peppers, tomatoes, etc.) were insufficient for household needs.[15]

Since the early years of the revolution there have been food shortages in Cuba. In the early 1960s, the need for the rationing system came about due to an increase in demand for food as Cubans gained purchasing power and a decrease in domestic food production resulting from the shift toward state ownership of farmland and food production enterprises[16] (Alvarez 2004). Cuba's national food rationing system has been in place since 1962. It is referred to locally as "*la libreta*," a reference to the physical ration card. Every Cuban is eligible for a ration card with which they can purchase basic food items. Prices are very heavily subsidized. The food items in the monthly ration at the time of my research included: five to ten eggs, five pounds refined sugar, five pounds raw sugar, five to ten pounds of white rice, five pounds of beans, 0.4 pints cooking oil, a roll of bread per day, and 200g-500g pork or ground beef mixed with soy.[17] All of these items cost about 25 national pesos a month, the equivalent of about one U.S. dollar [at the official exchange rate]. The ration system was initially intended to provide many more food items, but the food shortages common in Cuba since the 1960s have limited what the state has been able to provide in the ration.

While the rations are the most common and most important way that Santiagueros acquire food, all of the people I interviewed claimed that no one would be able to survive on rations alone. Most participants stated that the food provided in the ration only lasted from one to two weeks rather than the full month. Santiagueros supplement their monthly food rations through the alternative ways of acquiring food, listed here in order of importance: peso purchases, CUC *chopin*,[18] black market purchases, and informal trades, self-production, and gifts.

Acquiring food outside the rationing system

Paradoxically, many Santiagueros said that the transition to state control over food production in the 1960s caused a great shift in the everyday eating of a typical Cuban household, yet when asked what typical Cuban meals consisted of before the revolution, most responded that the standard foods – rice, black beans, and tubers – had in fact not changed. It is true, however, that for many Santiagueros the acquisition of these ingredients has become increasingly difficult. The insistence on maintaining essentialized fare despite mounting difficulties in food acquisition illustrates the centrality of these food traditions in Santiagueros' lives.

After the collapse of the Soviet Union, Cubans experienced the worst economic crisis since the 1959 revolution, a period the Cuban government calls the "Special Period in Time of Peace" (Deere 1995; Gaceta 1993; Messina 2004; Ritter 1990; Stricker 2007). The end of the Cold War left Cuba without major allies for economic trade; from 1989 to 1992, imports were cut by 73 percent, from $8.1 billion to $2.2 billion (Preeg 1993; Roca 1977). With the drastic reduction of oil, fertilizer, and pesticide imports from the Soviet

Union, the agricultural crisis gave rise to food emergency – the bread ration was cut by twenty percent in Havana to three ounces daily and the bread price was raised by thirty percent outside of Havana (Chaplowe 1996). The tightening of the U.S. embargo and the caps on remittances from Cuban Americans created further complications, which at the time of this research still had a significant impact on the everyday lives of Cubans on the island. Nearly twenty years after the start of the Special Period, while food production and availability have improved, there is still need for further improvement (Ritter 2004). The day-to-day efforts that Cubans employ in order to acquire food and other goods is essential to making the system work; food acquisition is a daily, interdependent activity bringing people together as a community. Through social networks Santiagueros work tirelessly within and outside of legal boundaries in order to be able to produce what they see as a "decent meal," not just for their families but for the families of friends, neighbors, coworkers and others alike (DeVault 1997; Weismantel 1989).

Cuba has a dual currency economy. The CUC is a close equivalent to the Euro, and the national peso is worth about four cents. Cubans get paid in the national peso. There are many ways to make food purchases with the Cuban national peso (CUP). At the time of this research there were five large markets that sell goods only for pesos in Santiago. This type of market was originally called the Parallel Market – *mercado paralelo* – when it was established in the 1970s; however, now most people refer to these markets simply as *the* market – *el Mercado* or *Los Mercados*. Many of the people that I interviewed noted that these markets are very expensive for them, so much so that they try to only shop at these markets when absolutely necessary, that is, once they have consumed everything provided by the ration and any foods they can acquire in cheaper ways.[19] These markets sell many items that are not available through the rations, such as fresh fruits and vegetables, and they also sell items that are in the ration, such as rice, beans, coffee, and other dry goods, thus allowing families to acquire more of these foods when the ration is not sufficient. While these markets often have inconsistent supplies, they are often the only places where certain types of food items are available, such as fruits and vegetables that easily perish or are not conducive to sale by street vendors. Those with more resources use these markets more, but complain of their limited and irregular supply. Some of these items are also available through street vendors; however, due to the small quantity they are able to carry, they often run out of popular items. There are also food stalls that are heavily subsidized by the Cuban government where people often buy food. However, these stalls are not usually well supplied. In addition to peso markets, street vendors, whether they have a permit to legally sell food or not, accept pesos for their goods. Food sales between friends and family normally use the national peso. Many black market purchases are made in pesos, though illicit vendors increasingly demand CUC for their products.

Items in the CUC market are priced similarly to U.S. prices (Garth 2009; Gordy 2006).[20] While there are many items available for sale in the CUC markets, they are accessible to the few Santiagueros who have extra monthly income. Most of the Santiagueros that I interviewed were able to make very few CUC purchases each month. Many cited soap and deodorant as their most important monthly CUC purchases, and very few people mentioned purchasing the food items available in the CUC stores. Those with access to remittances or other means of acquiring CUC are likely to make more of these types of purchases.

As the amount of food in the ration has decreased and the number of foods available only in CUC increased, many told me that they turn to the black market (see also del Real and Pertierra 2008). Although unauthorized beef consumption and/or cattle slaughter is a punishable crime, my informants reported that black market beef is probably the most sought after black market food product in all of Cuba.[21] Many remember a time when beef was fundamental to Cuban fare and continue to express its centrality through reminiscing about traditional Cuban dishes such as *ropa vieja*. Frozen, imported lobster, shrimp, and other seafood destined for hotels and resorts are also traded on the black market. These foods, along with myriad other items that originate in the hospitality industry, from industrial-sized cans of vegetables to foreign beer, constitute a large proportion of the black market goods.[22]

The process of acquiring food in Santiago de Cuba is very time consuming for most households; often there is one family member whose full-time efforts are dedicated to food acquisition and preparation. The food scarcities, long lines, and the dual currency system all exacerbate a process that was already very time consuming given that there are so many different markets and locations to travel to for food acquisition. As the case of Georber illustrates below, all of these difficulties of food acquisition result in Santiagueros characterizing the process as a *lucha* or part of *la lucha* - a daily struggle to acquire food.

"Disconnecting the mind" and the struggle to acquire food

On one of Santiago's hottest summer days I arrived on time for my third interview with Georber, a 55-year-old, mestizo, financially well-off musician, to find a note on his door requesting that I wait for him at the neighbor's house. Georber works independently as a musician; he does not receive pay from the state for his work but relies solely on the direct sales of his recordings and pay for performances. He still receives the state social benefits, including the ration card. He has traveled throughout Europe performing his music and has been able to save enough money to live very well in Santiago. He lives in a large three-bedroom house that he has all to himself. He is unmarried with no children. His brother, whom he supports financially, lives next door. After I had chatted with the neighbors on their balcony for two hours, Georber rode up on a *moto* with several canvas bags filled with food in his hand. He seemed exhausted when he came to fetch me, and after apologizing, he explained to me that he had to travel to the outskirts of town to get the *yuca* that he needed for lunch that day. The markets in town only had *malanga (Xanthosoma)*, which he is prohibited from eating for religious reasons, and he needed *yuca* for a particular dish. He told me that while having to spend two hours going from market to market in search of food is quite common, he was stressed about it today because he was unsure if I, as a foreigner, would understand and wait for him. He then caught himself and said, "Oh, clearly you understand, you are studying food!" He had to pause and smoke a few cigarettes before we could start our interview in order to calm down after his stressful morning. While he was trying to relax he explained that for him it is essential to "leave his mind at home" when he goes out shopping for food, that without detaching his mind, there were just too many things to worry about. He added that "if I had left my mind at home I would not be damaging my health right now with the stress of not being able to find the exact food that I wanted and I would not have to smoke so many cigarettes to calm down." Georber's frustration with the scarcity of *yuca* is indicative of a more general sentiment in Santiago that people should be able to access the foods that they need, ingredients for essentialized fare, conveniently and affordably. For him, disconnecting the mind is an act used to alleviate mental distress and to relieve many corporeal aspects of stress as well.

While not everyone made references to "disconnecting the mind," almost everyone over the age of 25 that I spoke to repeatedly made reference to the practice in various ways, including: recounting memories of difficult situations that they have to still "disconnect the mind" from; giving me the advice to "disconnect my mind" from my own frustrations; or by simply commenting *"hay que desconectar la mente"* (one has to disconnect the mind). Additionally, I found that when people were reflecting on the act of "disconnecting the mind," their emotional responses were quite varied but always extremely intense, often pulling me into their emotional state. Most participants shifted their tone and affective state when referring to the act of "disconnecting the mind"; they spoke more slowly, in a less determined manner, and appeared to be intently reflective. It was this shift in voice that initially drew my attention to the importance of this concept. Conversations about "disconnecting the mind" almost always lacked the black humor that is so common to Cuban conversations about the struggles of everyday life on the island. Both men and women cried or became teary-eyed when talking about the need to disconnect the mind, while others spoke of it as if it were an act of liberation. Similarly, the refrain, "Ay, it's not easy" is a common Cuban expression not only of the day-to-day struggles to maintain Santiago life, but also of the struggle to remain disconnected from the *la lucha* in order to persevere. "Disconnecting the mind" is directly related to the stresses and difficulties derived from *la lucha*. In addition to its use as a mechanism for coping with mentally stressful situations, it appears that "disconnecting the mind" is an idiom used to describe the process through which one can escape the toils of these daily struggles and maintain a pleasant disposition.

Efforts to "disconnect the mind" are closely tied to memories of traditional cuisine and articulations of essentialized fare. For instance, María Isabel explained to me that before she leaves the house in the mornings to begin her daily food shopping, she sits down to enjoy a cup of Cuban coffee. It is through this morning ritual that she claims she is able to "disconnect her mind from her body," something that she feels she must do in order to face the daily struggles of food acquisition and still be able to maintain a personable, pleasant disposition as she makes her rounds to the various places where she gets the products that she needs for the day. The scarcity of food and consumables in general is of ongoing concern for Santiagueros, who tend to receive even fewer products than people in Havana. She explained that if she did not "disconnect her mind":

> the stress and anxiety of getting food every day would just cause too much worry. I wouldn't be able to converse with my neighbors, my friends, family. I would just worry and be inside my head with my thoughts all the time. I would just worry and the stress would be bad for me, it would kill me.

For María Isabel, problems with food availability and the effective devaluation of the national peso in Cuba are interpreted as part of the trials and tribulations of living in a socialist society, something she actually values very highly. It is important to point out that María Isabel has participated in several of Cuba's medical aid missions, where Cuban health professionals are sent to foreign countries that are in need of medical personnel.[23] María Isabel's pride in her service work for her country, and the dire poverty that she witnessed in other countries, directly connect to her shame and frustration with the process of acquiring the foods necessary to be able to serve her family a decent meal.[24] As described earlier, she characterizes this as a *"golpe psicológico."*[25] Through these words, she expresses the difficulties that she encounters in making the system work. She believes that she must disconnect in order to deal with this psychological blow regarding the discrepancy between her sense of nationalism and pride in her country and her frustration with the struggle of everyday life there.

According to another interviewee, Mercedes, the food shortages have always been a problem, and those of the "Special Period" have not ended. When I asked her how long ago her family's food problems had started, tears came to her eyes, her voice faltered, and she said:

> We now have about thirty years of "Special Period" . . . Imagine it, adapting yourself to it, here everything had to be planned. But, also, at the start of this there used to be special markets, called parallel markets, where there were products for open sale. You had access, paying the price that they wanted. If your rice wasn't enough you went there and you bought rice, oil, like now with the CUC stores. But not everyone has access to the CUC, here no one makes [earns] CUC.

Here, Mercedes reflects on the fact that she and her family have struggled with food scarcity issues for thirty years, when in fact the state's officially designated "Special Period" has been in place for just under twenty years. Her statement about the "thirty years of 'Special Period'" reveals her perspective that although the scarcities worsened during the 1990s, they have long been a source of struggle for Santiagueros. For her, this struggle is more than simply overcoming food scarcity, but the struggle to adjust to the constant changes in state policies and state controls over food access that, in turn, becomes a source of stress for Mercedes. By noting the introduction of CUC and inequalities in access to CUC, she is pointing out another state policy that has made it more difficult for her and her family to access foods and other goods. She reflects on the increasing distinction between the lifestyles of those who have CUC and those who do not. The need for "everything to be planned" and constantly not knowing what the next changes in the system will be create uncertainty and feelings of insecurity that cause stress, which is articulated locally as part of the struggle of everyday life.

Lines at the currency conversion points, the *cadeca*, often wind around the block, giving Santiagueros like Carlos plenty of time in the hot sun to contemplate the allocation of these precious CUC. In an interview, Carlos, a middle-class Santiaguero of Chinese descent, pointed out to me that to really understand this situation you have to look at Cubans who don't get remittances, or have any other way of getting CUC:

> How DO those people survive, I don't know how, I don't understand. There are people who only eat one time a day and I don't think that is good. While it's true that here no one dies of hunger, because people find a way, they invent something so that they can eat, but if you invest in one thing you can't invest in another.[26]

Carlos expresses his own anxiety about the situations of other Cubans who have even less than he. He worries, for instance, about how those who must choose between eating and clothing their children will figure out how to get by. After asking Carlos to tell me about typical meals that he and his first wife shared at the beginning of their relationship, Carlos spoke about how scarcity issues still continue today:

> The problem is that really, let's say that the scarcity problems have always remained, in other words oil – there wasn't much – meat, rice, in other words nothing, things in the ration. . . One has to disconnect from this, you can't use too much energy in worrying about this.

Carlos's conclusion that "one has to disconnect" in order not to "use too much energy worrying" resonates with Rodulfo's description of the act of disconnecting the mind. In an interview I asked Rodulfo what he and his family do when they can't find the types of foods that they desire. Rodulfo, a black 63-year-old retired engineer who lives in a small Santiago home with his mother, three sisters, and eight nieces and nephews, responded "one has to disconnect the mind from those things." I prodded him to explain what he meant by this:

> We live in a country where there is always something that is lacking, for every generation there is something that we desire but that we can't have access to. If someone worries about how they will get things when they can't get things they will go crazy, the nerves, the anxiety, the stress will drive them crazy. All of that stress is bad for your health, your bodily health, and it's bad for your community, for your family. That is why you have to disconnect your mind from those things. People who don't disconnect their minds will never overcome because there is always something that's lacking and they will make everyone around them sad and angry, they will lose their lives, lose anything worth living for. You have to disconnect in order to connect with life, in order to continue living.[27]

As Rodulfo clearly states, "disconnecting the mind" is an act with a long and rich history in Cuba, although as the sources of difficulties of life have changed throughout Cuban history the struggles that cause one to have to "disconnect" have shifted as well.[28] Since 1959, Cubans have "disconnected" from the various struggles related to the dramatic political changes of the revolution, including separation from family members, loss of property and businesses, and drastic shifts in the processes of acquiring the goods necessary for daily life. It appears that in recent years, particularly in Santiago, food scarcity has been a major source of struggle that often leads to the need to "disconnect the mind."

Discussion and Conclusion

The data presented here provide a glimpse into the struggle of daily food acquisition in Santiago de Cuba, a struggle that manifests not only economically and politically but also socially and psychologically. When constraints prevent access to certain goods, in this case foods, people cope with constraint in various ways. Some change some food habits, substituting ingredients when acceptable and creating new consumption situations. In this particular case, rather than changing food consumption patterns, Santiagueros engage in a struggle to maintain their way of eating, a struggle that they later cope with by "disconnecting their minds" from it. Focusing on the local idiom "disconnecting the mind," this chapter illustrates the ways in which Santiagueros cope with difficulties associated with food consumption and acquisition and the social costs of maintaining one highly valued way of eating despite mounting difficulties.

The struggle to assemble what I call "essentialized fare" is imbued with local and national histories of political and social structures. Essentialized fare is the local notion that Cubans "were made to eat" certain kinds of foods, or that all Cubans currently have and always have had certain taste preferences. Notions of essentialized fare are deeply tied to taste, desire, and local ideas of what is considered to be appropriate to eat. One such example is the ubiquitous statement: "Cubans like to eat meat, not vegetables," which I encountered in nearly all of my interviews, but not always in my observations. That is to say, the term "essentialization" is used here to show that these alimentary practices are discursively essentialized or thought to be an inherent part of Cuban identity, yet in practice individual and family behavior does not necessarily align with these essentialized ideals. The consumption of essentialized fare may be an ideal that is often not met in practice; indeed essentialized fare is also idealized fare. The essentialization here is a reduction of more complex sociopolitical phenomena related to longing for progress without all of the difficulties of the current system for acquiring goods, which results in a tension between longing for "traditionality" (as Cubans put it) and a simpler way to acquire goods, on one hand, and the desire for progress, still associated with socialist ideals, on the other. Additionally, as Martin's interview suggests, new types of food and changes in Santiago dining are increasingly converted to forms of symbolic capital, signaling certain kinds of prestige. Still, for Martin there is a large degree of symbolic and cultural capital in the maintenance of essentialized fare.

The emotional response to the act of "disconnecting the mind" is closely tied to a general emotional disposition toward the struggle to maintain Cuban culinary and cultural identity. As the quotes from Martin and María Isabel above illustrate, there are several factors, including tourism, media, and changing government policies, that impact levels of satisfaction with the local food system. Expressing fear, anxiety, heightened stress, and the need to remain detached through "disconnecting the mind," these interlocutors have expressed the different ways in which they cope with the difficulties of accessing goods in their everyday lives. This analysis of the local significance of essentialized fare and the act of disconnecting the mind reveals the importance of analyzing tacit communication around sources of interpersonal, social, political, economic, and spiritual distress. While disconnecting the mind from the struggle to acquire essentialized fare may not be an idiom of distress that indicates a psychopathological state, it is indicative of a socially significant experience of distress and a possible indicator of feelings of anger, powerlessness, and insecurity.

For some, such as María Isabel, the fight for traditional food consumption is tied up with resistance to the current state food system. Insistence on essentialized fare becomes a form of protest. As María Isabel pleads that Santiagueros hold on to and maintain traditional dishes despite the difficulties of acquiring these foods, she expresses a strong resistance to the current food system. I believe that acts such as "disconnecting the mind" are particularly meaningful in the face of the emotional stresses that come with this situation. "Disconnecting the mind" is used as a way to cope with and resist the decreases in the quality and quantity of state provisioning, and to persevere and endure through a daily struggle to maintain a decent quality in spite of these changes. To disconnect with these struggles becomes a way for Santiagueros to insist on maintaining their quality of life and to connect with everyday life in the present.

As Biehl, Good, and Kleinman have shown in other settings, these subjects also "struggle with the possibilities and dangers of economic globalization, and the threat of endless violence and insecurity, and the new infrastructures and forms of political domination and resistance that lie in the shadows of grand claims of democratization and reform" (Biehl et al. 2007:1). Notions of Cuban essentialized fare and the Santiaguero ideology of a decent meal continue to haunt Santiagueros despite the fact that, as Mercedes notes, food shortages have been commonplace for over thirty years now. These food scarcities force Santiagueros to grapple with "what really matters" and what is at stake in leading an adequate life (Kleinman 2006). In this case, what really matters for many Santiagueros is the ability to maintain what they view as a decent standard of living through their food. Though the ideal meal has changed little, I have shown that the social processes of acquisition have indeed changed, and I posit that what the meal symbolically constitutes has also shifted greatly. The maintenance of essentialized fare is a symbolic way of holding on to imagined traditions despite mounting difficulty accessing foods in Santiago de Cuba.

The Santiaguero longing to maintain Cuban essentialized fare even at very high financial and social costs is a contemporary form of resistance to social fractures that have been intensified by recent political changes in Cuba. The desire to maintain essentialized fare is precisely the symbolic tool that is used to repair these fractures. In this context, the insistence on maintaining essentialized fare is a social practice that holds steady in the face of constant disappointments. The Santiagueros in this study express general frustration with the Cuban food system and particular frustrations with the difficulties of the daily struggle to make the system work. It is this struggle to survive, the struggle to fight against the many difficulties that affront them – as manifested by the concept of a "*golpe psicológico*" – that necessitates "disconnecting the mind." In turn, the act of "disconnecting the mind" allows Santiagueros to maintain continuity of their national identity despite the difficulty of maintaining a cuisine symbolic of an adequate quality of life.

References

Adler, S.R. (1995) 'Refugee stress and folk belief: Hmong sudden deaths', *Social Science and Medicine*, 40.12: 1623-1629.

Allen, M.W. and Baines, S. (2002) 'Manipulating the symbolic meaning of meat to encourage greater acceptance of fruits and vegetables and less proclivity for red and white meat', *Appetite*, 38: 118-130.

Alvarez, J. (2004) *Cuba's Agricultural Sector*, Gainesville, FL: University Press of Florida.

Anagnost, A. (1995) 'A surfeit of bodies: population and the rationality of the state in post-Mao China', in F. Ginsburg and R. Rapp (eds.) *Conceiving the New World Order: The Global Politics of Reproduction*, Berkeley, CA: University of California Press.

Andaya, E. (2009) 'The gift of health: socialist medical practice and shifting material and moral economies in post-Soviet Cuba', *Medical Anthropology Quarterly*, 23.4: 357-374.

Appadurai, A. (1988) 'How to make a national cuisine: cookbooks in contemporary India', *Comparative Studies in Society and History*, 30: 3-24.

Askegaard, S., Kjeldgaard, D. and Arnould, E.J. (1999) 'Identity and acculturation: the case of food consumption by Greenlanders in Denmark' (MAPP Working Paper No. 67), University of Aarhus: the MAPP Centre.

Barthes, R. (1997) 'Towards a psychosociology of contemporary food consumption', in C. Counihan and P. van Esterik (eds.) *Food and Culture: A Reader*, New York: Routledge.

Bell, D. (2003) 'Mythscapes: memory, mythology, and national identity',

British Journal of Sociology, 54.1: 63–81.

Benjamin, M., Collins, J. and Scott, M. (1984) *No Free Lunch: Food and Revolution in Cuba Today*, Princeton, NJ: Princeton University Press.

Bentley, A. (2001) 'Reading food riots: scarcity, abundance and national identity,' in P. Scholliers (ed.) *Food, Drink and Identity: Cooking, Eating and Drinking in Europe Since the Middle Ages*, Oxford: Berg Publishers.

Berdahl, D. (2005) 'The spirit of capitalism the boundaries of citizenship in post-wall Germany', *Comparative Studies of Society and History* 47.2: 235-251.

Biehl, J., Good, B. and Kleinman, A. (2007) *Subjectivity: Ethnographic Investigations*, Berkeley and Los Angeles, CA: University of California Press.

Bott, E. (1987) 'The Kava ceremony as a dream structure,' in M. Douglas (ed.) *Constructive Drinking: Perspectives on Drink from Anthropology*, Cambridge: University of Cambridge Press.

Bourdieu, P. (1984) *Distinction: A Social Critique of the Judgment of Taste*, Cambridge, MA: Harvard University Press.

— (1993) *The Field of Cultural Production: Essays on Art and Literature;* trans R. Johnson, New York: Columbia University Press.

Brodwin, P. (2002) 'Genetics, identity, and the anthropology of essentialism', *Anthropological Quarterly*, 75.2: 323-330.

Brotherton, P.S. (2005) 'Macroeconomic change and the biopolitics of health in Cuba's Special Period', *Journal of Latin American Studies*, 10.2: 339-369.

Cabezas, A. (2009) *Economies of Desire: Tourism and Sex in Cuba and the Dominican Republic*, Philadelphia, PA: Temple University Press.

Caldwell, M. L. (2002) 'The taste of nationalism: food politics in postsocialist Moscow', *Ethnos*, 67.3: 295-319.

— (2004) 'Domesticating the french fry: McDonald's and consumerism in Moscow', *Journal of Consumer Culture*, 4.5: 5-26.

Carrier, J. (2006) 'The limits of culture: political economy and the anthropology of consumption', in F. Trentmann (ed.) *The Making of the Consumer: Knowledge, Power, and Identity in the Modern World*, Oxford: Berg.

Castro, F. (1967) *History Will Absolve Me*. Havana: Guairas.

Chaplowe, S.G. (1996) *Havana's Popular Gardens and the Cuban Food Crisis*, master's degree thesis, University of California, Los Angeles.

Counihan, C.M. and Kaplan, S.L. (eds.) (2004) *Food and Gender: Identity and Power*. London: Routlege.

Dawdy, SL. (2002) 'La comida mambisa: food, farming, and Cuban identity,

1839–1999', *New West Indian Guide*, 76: 47-80.

Davis, B. and Tarasuk, V. (1992) 'Hunger in Canada', *Agriculture and Human Values*, 11.4: 50-57.

de Certeau, M. (1984) *The Practice of Everyday Life*. Berkeley, CA: University of California Press.

De La Torre, M.A. (2003) La Lucha for Cuba: Religion and Politics on the Streets of Miami, Berkeley and Los Angeles, CA: University of California Press.

Deere, C.D. (1995) 'The new agrarian reforms', NACLA Report on the Americas Special Issue on Cuba: Adapting to a Post-Soviet World, 29.2: 13-17.

del Real, P. and A.C. Pertierra (2008) 'Inventar: recent struggles and inventions in housing in two Cuban cities,' *Buildings and Landscapes*, 15: 78-92.

Derby, L. (1998) 'Gringo chickens with worms', in G. Joseph, C. LeGrand and R. Salvatore (eds.) *Close Encounters of Empire: Writing the Cultural History of US Latin American Relations*, Durham, NC: Duke University Press.

Desjarlais, R., Eisenberg, L., Good, B. and Kleinman, A. (1995) *World Mental Health: Problems and Priorities in Low-Income Countries*, New York and Oxford: Oxford University Press.

DeVault, M. (1997) 'Conflict and defense', in C. Counihan and P. van Esterik (eds.) *Food and Culture: A Reader*, New York: Routledge.

Douglas, M. and Isherwood, B. (1979) *The World of Goods*, London: Penguin Books

Espina Prieto, M. (2004) 'Social effects of economic adjustment: equality, inequality, and trends toward greater complexity in Cuban society', in J. Dominguez, O.E. Perez Villanueva, L. Barberia (eds.) *The Cuban Economy at the Start of the Twenty-First Century*, Cambridge, MA: Harvard University Press.

Feinsilver, J. (1993) *Healing the Masses: Cuban Health Politics at Home and Abroad*, Berkeley, CA: University of California Press.

Gaceta Oficial (1993) in *Gaceta Oficial de la República de Cuba*, La Habana.

Gal, S., and Kligman, G. (2000) *The Politics of Gender after Socialism: A Comparative-Historical Essay*, Princeton, NJ: Princeton University Press.

Garth, H. (2009) 'Things became scarce: food availability and accessibility in Santiago de Cuba then and now', *NAPA Bulletin*, 32.1: 178-192.

Garth, H. (ed.) (2013) *Food and Identity in the Caribbean*, London: Berg

Publishers.

Gofton, L. (1986) 'The rules of the table', in C. Ritson, J. McKenzie and L. Gofton (eds.) *The Food Consumer*, New York: John Wiley and Sons.

Gordy, K. (2006) "'Sales + economy + efficiency=revolution?" dollarization, consumer capitalism, and popular responses in Special Period Cuba,' *Public Culture*, 18.2: 383-412.

Hadjiyanni, T. and Helle, K. (2009) 'Re/claiming the past--constructing Oijbwe identity in Minnesota homes', *Design Studies*, 30.4: 462-481.

Hall, S. (1996)   'New ethnicities', in H.A. Baker, M. Diawara and R.H. Lindebord (eds.) *Black British Cultural Studies: A Reader*, Chicago, IL: University of Chicago Press.

Hastrup, K. (1993) 'Hunger and the hardness of facts', *Man: Journal of the Royal Anthropological Institute*, 28: 727-739.

Hernandez, R. (2008) Interview with Rafael Hernandez, Director of Temas: Democracy in Americas.

Hinton, D.E., Pich, V., Marques, L., Nickerson, A. and Pollack, M.H. (2010) 'Khyâl attacks: a key idiom of distress among traumatized Cambodia refugees', *Culture, Medicine, and Psychiatry*, 34.2.

Holgado Fernández, I. (2000) *¡No es Facil! Mujeres Cubanas y la Crisis Revolucionaria*, Barcelona: Icaria.

Holtzman, J. (2003) 'In a cup of tea: commodities and history among Samburu pastoralists in northern Kenya', *American Ethnologist*, 30.1:136-155.

Kaplan, M. (2007) 'Fijian water in Fiji: local politics and a global community', *Cultural Anthropology*, 22.4: 685-706.

Kahn, M. (1986), *Always Hungry, Never Greedy: Food and the Expression of Gender in a Melanesian Society*, London and New York: Cambridge University Press.

Khan, A. (2004) *Callaloo Nation: Metaphors of Race and Religious Identity among South Asians in Trinidad*, Durham, NC: Duke University Press.

Kirmayer, L. J. and Young, A. (1998) 'Culture and somatization: clinical, epidemiological, and ethnographic perspectives', *Psychosomatic Medicine* 60: 420-430.

Kleinman A. (2006) *What Really Matters: Living a Moral Life Amidst Uncertainty and Danger*, Oxford: Oxford University Press.

Kligman, G. (1998) *The Politics of Duplicity: Controlling Reproduction in Ceausescu's Romania*, Berkeley, CA: University of California Press.

Klumbyté, N. (2010) 'The Soviet sausage renaissance', *American Anthropologist*,

112.1: 22-38.

Kohrt, B. A. and Hruschka, D.J. (2010) 'Nepali concepts of psychological trauma: the role of idioms of distress, ethnopsychology and ethnophysiology in alleviating suffering and preventing stigma', *Culture, Medicine and Psychiatry*, 34.2.

Lee, D.T.S., Kleinman, J. and Kleinman, A. (2007) 'Rethinking depression: an ethnographic study of the experiences of depression among Chinese', *Harvard Review of Psychiatry*, 15: 1–8.

Lewis-Fernández, R., Gorritz, M., Raggio, G.A., Peláez, C., Chen, H and Guarnaccia, P.J. (2010) 'Association of trauma-related disorders and dissociation with four idioms of distress among Latino psychiatric outpatients', *Culture, Medicine and Psychiatry*, 34.2.

Mesa-Lago, C. (2005) 'The Cuban economy in 2004-2005', in *Cuba in Transition* (ASCE and ICCAS), 15: 1-18.

Messina, W.A. (2004) 'Cuban agriculture in transition: the impacts of policy changes on agriculture production, food markets and trade', in A.R. Ritter (ed.) *The Cuban Economy*, Pittsburgh, PA: University of Pittsburgh Press.

Miller, D. (ed.) (1995) *Worlds Apart: Modernity through the Prism of the Local*, London: Routledge.

Mintz, S.W. (1985) *Sweetness and Power: The Place of Sugar in Modern History*, New York: Viking.

— (1996) *Tasting Food, Tasting Freedom: Excursions into Eating, Culture, and the Past*, Boston, MA: Beacon Press.

Nichter, M. (1981) 'Idioms of distress: alternatives in the expression of psychosocial distress: a case study from South India', *Culture, Medicine and Psychiatry*, 5.4: 379–408.

— (2010) 'Idioms of distress revisited', *Culture, Medicine and Psychiatry*, 34.2: 401–416.

Obeyesekere, G. (1981) *Medusa's Hair*, Chicago, IL: University of Chicago Press.

Pérez, L.A. Jr. (2009) 'Thinking back on Cuba's future: the logic of patria', *NACLA Report on the Americas*, March/April: 12-20.

Pertierra, A.C. (2007) 'Cuba: the struggle for consumption', unpublished doctoral thesis, University College London.

— (2008) 'En casa: women and households in post-Soviet Cuba,' *Journal of Latin American Studies*, 40: 743-767.

Pottier, J. (1999) *The Anthropology of Food: the Social Dynamics of Food Security*,

London: Polity Press.

Preeg, E. II and Levine, J.D. (1993) 'Cuba and the new Caribbean economic order', in *Significant Issues Series Vol. XV*, Washington DC: the Center for Strategic and International Studies.

Premat, A. (1998) 'Feeding the self and cultivation of identities in Havana Cuba', doctoral thesis, York University.

Ritter, A.R. (1990) 'The Cuban economy in the 1990s: external challenges and policy imperatives', *Journal of Interamerican and World Affairs*, 32.3: 117-50.

— (2004) 'The Cuban economy in the twenty-first century: recuperation or relapse', in A.R. Ritter (ed.) *The Cuban Economy*, Pittsburgh, PA: University of Pittsburgh Press.

Roca, S.G. (1977) 'Cuban economic policy in the 1970s: the trodden paths', in I.L. Horowitz (ed.) *Cuban Communism*, New Brunswick, NJ: Transaction Books.

Roseberry, W. (1996) 'The rise of yuppie coffees and the reimagination of class in the United States', *American Anthropologist*, 98.4: 762-775.

Rouse, C. and Hoskins, J. (2004) 'Purity, soul food, and Islam: explorations at the intersection of consumption and resistance', *Cultural Anthropology*, 19: 2.

Sahlins, M. (1990) 'Food as symbolic code', in S. Seidman (ed.) *Culture and Society: Contemporary Debates*, Cambridge: Cambridge University Press.

Scheper-Hughes, N. (1993) *Death without Weeping: The Violence of Everyday Life in Brazil*, Berkeley, CA: University of California Press.

Scholliers, P. (2001) *Food, Drink and Identity: Cooking, Eating and Drinking in Europe since the Middle Ages*, Oxford: Berg Press.

Stricker, P. (2007) *Toward a Culture of Nature: Environmental Policy and Sustainable Development in Cuba*, New York: Lexington Books.

Sutton, D. (2001) *Remembrance of Repasts: An Anthropology of Food and Memory*, Oxford: Berg Press.

Tanuma, S. (2007) 'Post-Utopian Irony: Cuban Narratives during the "Special Period" Decade', *PoLAR: Political and Legal Anthropology Review*, 30.1: 46-66.

Throop, C.J. (2003) 'On crafting a cultural mind: a comparative assessment of some recent theories of "internalization" in psychological anthropology', *Transcultural Psychiatry*, 40.1: 109-139.

USDA (2008) *Cuba's Food and Agriculture Situation Report*, Office of Global Analysis, FAS. Online. Available HTTP: <http://www.fas.usda.gov/itp/

cuba/cubasituation0308.pdf> (accessed 15 February 2013).

Weismantel, M.J. (1989) *Food, Gender, and Poverty in the Ecuadorian Andes*, Illinois: Waveland Press.

Whitehead, H. (2000) *Food Rules: Hunting, Sharing, and Tabooing Game in Papua New Guinea*, Ann Arbor, MI: University of Michigan Press.

Wilk, R. (1999) '"Real Belizean food": Building Local Identity in the Transnational Caribbean', *American Anthropologist*, 101.2: 244–255.

— (2006) *Fast Food/Slow Food: The Cultural Economy of the Global Food System*, Walnut Creek: Altamira Press.

Williams, R. (1977) *Marxism and Literature*, London and New York: Oxford University Press.

Wilson, M. (2009) '¡No tenemos viandas! Cultural ideas of scarcity and need', *International Journal of Cuban Studies*, 2.1: 1-9.

— (2010) 'Food as a good versus food as a commodity: contradictions between state and market in Tuta, Cuba', *Journal of the Anthropological Society of Oxford*, 1.1: 25-51.

Endnotes

1   I would like to acknowledge my friends in Santiago who have made this project possible by sharing their stories with me. In particular I thank the Casa del Caribe for supporting my work along with my doctoral committee members: Carole Browner, Linda Garro, Jason Throop, Akhil Gupta, and Robin Derby. I am grateful to Mara Buchbinder, Hadi N. Deeb, Christel Miller, Keith Murphy, Eve Tulbert, and Kristin Yarris for comments on earlier drafts. I would also like to thank the editor, Nancy Burke, and anonymous reviewers for their comments, which greatly improved this chapter. This project was generously funded by several organizations, including the Social Science Research Council, the National Science Foundation, the UC Cuba Initiative, the UCLA Latin American Institute, the UC DIGSSS program, the UCLA Department of Anthropology, the UCLA Center for Study of Women, and the UCLA Eugene and Cota Robles Fellowship. Similar versions of this paper were presented at the American Anthropological Association Annual Meeting in 2007 and received the Association for the Study of Food and Society Graduate Student Paper Prize in 2010.

2   This name and all of the other proper names in this article are pseudonyms.

3   *La lucha* has also been used to refer to the Cuban fight against the U.S.-backed

Machado and Batista regimes. In Miami, the term *la lucha* refers to the struggles of the exile community against the revolutionary government and the struggle to regain the Cuba that they feel they have lost (cf. De La Torre 2003).

4   For further discussion of idioms of distress, see Hinton et al. 2010; Kohrt and Hruschka 2010; Lewis-Fernández et al 2010; Nichter 1981, 2010.

5   According to Nichter (2010), there has been some critique of the notion of idioms of distress. For instance, "Kirmayer and Young (1998) have written that 'the notion of "idiom of distress" may be misleading, to the extent that such "idioms" are assumed to be highly structured and entirely conventional ways of expressing distress. In reality, the meanings expressed through these idioms are often fragmentary, tentative, and even contradictory.'" (Nichter 2010: 408). Furthermore, "Lee et al. (2007) have drawn attention to the way in which labeling embodied emotional expressions as metaphors or idioms leads us to think of them as figurative, and not genuine, and to valorize meaning above the lived experience of sensations linked to emotional states" (Nichter 2010: 408).

6   The notion of "Afro-Cuban" is largely a concept created and sustained by foreigners conducting research in Cuba; Cubans themselves rarely use this term (Brotherton 2005). Rather than identify as Afro-Cuban, most Cubans of African descent simply identify as Cuban.

7   Cubans today often have ideas of what life was like in the past on the island that may not be historically accurate. Nonetheless, in this article I honor subjects' own interpretations and understandings of the past because it is through these notions that people develop their current sense of identity (cf Bell 2003; Brodwin 2002; Sutton 2001).

8   Discourse essentializing Cuban taste and characteristics of *Cubanidad* might lead the undiscerning ear to assume a general acceptance of what it means to be Cuban and part of *"el pueblo Cubano."* But while Santiagueros often elicit tropes of Cuban national identity, many are also quick to distinguish themselves regionally, ethnically, linguistically, and along class lines.

9   For an elaboration on this joke, see Cabezas 2009 and Wilson 2009.

10   While many research participants commented about Cuban longings for foods that were once available on the island and that they conceive of as abundant in foreign countries, I observed very few households that actually tried to prepare "foreign foods" at home.

11   I should note while many participants conceive of foods such as Chinese food, pizza, and spaghetti as "new" to Cuba, these food items have a long history on the island via migrations to and from Cuba in the 19th, 20th, and 21st centuries. This category of "foreign food" is distinguished from many other forms of

foreign or newly introduced foods such as mass-produced packaged products.

12   I would like to thank an anonymous reviewer for pointing out that the tension between "traditional" foods and cheaper convenience foods is further complicated by the history of state promotion of certain foods during periods of scarcity. For instance, in the 1960s government pizzerias were established due to scarcity of ingredients. Here, Martin comments on the present-day meaning of foods such as pizza and Chinese food, which for himself and his family are related to what he observes as an increasing desire for the Euro-American aesthetic. While his own perception of these types of foods is derived from his present-day living situation, the meaning of these foods within his community may be much more complicated given the historical uses of these foods by the state.

13   Most Cuban dishes, both fried and non-fried, are made with at least some cooking oil; many subjects lamented that the cooking oil that they had access to was grossly insufficient for their needs.

14   Culantro (*Eryngium foetidum*) is a commonly used herb in Cuban cooking.

15   The scarcity of fresh vegetables in Cuba since the collectivization of state agricultural production is due in part to the inefficiency of the volunteer labor brigades used to harvest foods from state farms. These foods require high labor inputs and are often difficult to process and transport. Therefore, many of these food items would either spoil before reaching the population or were pilfered by the volunteer laborers. In the 1980s, "parallel markets" were created to attempt to alleviate the problem of access to fresh vegetables [See my comment with Mercedes above]. In these markets the farmers could set the prices for the goods that they sold to the population; farmers began making a great deal of money through these markets. Soon the state closed these markets due to the fact that the vendors were making too much money by "exploiting" the population. While these markets were expensive for many Santiagueros, the availability of quality food products was rejoiced upon and the loss of this system is often lamented when subjects compare it to food availability today. See also Garth 2009.

16   Cuba's post-revolutionary government implemented agrarian reforms in 1959 and 1962, which eliminated a land ownership system through redistribution and nationalization of the land and encouraged the development of cooperatives on large estates. In an effort to improve food access, the post-revolutionary government encouraged the cultivation of easily grown, highly nutritive crops such as *viandas* (a category that includes tubers such as *yuca*, *malanga* sweet potato, and yam, as well as plantains). However, the agricultural transition has not

resulted in sufficient food production to feed the population. Today more than 80% of Cuba's food is imported (USDA, 2008).

17   Also, children up to seven years old get one liter of milk daily, and children from seven to 14 years old receive one liter of yogurt daily.

18   When making purchases in CUC, Santiagueros use the word *chopin*, from the English word "shopping," as both a verb to describe the act of making CUC purchases and as a noun to describe the location in which the purchases are made. Statements such as "I am going *chopin*" and "I am going to the *chopin*" are used interchangeably to denote the act of making CUC purchases. Despite the discursive distinction between CUC *chopin* and other types of purchases, currently *chopin* in the CUC stores is the only way to acquire many of the products Cubans feel are necessary, including additional cooking oil, imported spices, and personal hygiene products such as deodorant.

19   The prices at these markets vary greatly. During the summer of 2008, I calculated the average prices of some foods: okra five pesos per piece (0.23 USD), small cucumbers three pesos (0.14 USD), small slices of squash one peso (0.05 USD), *malanga* nine pesos per pound (0.41 USD), *yuca* 2.50 per pound (0.11 USD), small plantains 7.50 per pound (0.34 USD), and tomatoes four pesos per pound (0.18 USD).

20   Note that after converting to CUC, a Cuban worker who receives 300 national pesos a month in salary has only 12 CUC a month.

21   Beef has been extremely scarce, heavily policed, and in turn highly sought since the mid-1960s. Cuban discourse about beef consumption is tied up with political resistance to the economic laws in the 1960s and the mismanagement of collective farms. Cuban beef and dairy production never fully recovered; locally produced beef and dairy products are still extremely scarce in Cuba. In 1962 a law imposing an age limit of seven years on access to milk rations was implemented. In 1964, a law mandated an eight-year prison sentence for killing, eating, or selling beef cattle in Cuba. (cf. Dawdy 1998; Mesa-Lago 2005).

22   I also learned that the general scarcity of goods and the extreme irregularity of product availability in Santiago lead to the hoarding of goods during times of abundance and subsequent trading and gift giving within social networks. Those with more resources buy products in bulk when available; later they sell or trade these items to friends, family, and extended social networks. For instance, Mickey, who runs a local restaurant and thus has increased access to CUC, circulated food gifts on a daily basis. While out running daily errands, Mickey always stops at his mother and grandmother's houses, with a quarter of a chicken for each one. Since there is seldom chicken in the ration and it's much more expensive

than pork in the markets, Mickey is able to provide gifts for his family that they cannot access themselves.

23 Since the 1959 revolution, Cuba's international medical and public health missions have been central to official narratives of anti-imperialism and the moral victory of socialism over capitalism (see Andaya 2009; Castro 1967; Feinsilver 1993).

24 For further discussion of these issues, see Mayra Espina Prieto's chapter "Social Effects of Economic Adjustment: Equality, Inequality, and Trends Toward Greater Complexity in Cuban Society" in Dominguez et al. 2004.

25 The word *golpe* in Spanish can mean "a blow" as in a punch, a knock, or a crash. It can also mean "suddenly" or "all at once," and it can mean a "coup" as well.

26 Cubans often use the expression "to invent something" to signify turning to the black or gray market or other less conventional means of acquiring what is needed. For further discussion of the Cuban concept of *inventar* see del Real and Pertierra 2008.

27 I asked Rodulfo if "disconnecting the mind" was related to the less common Cuban saying, "disconnect from the system" (*desconectar del sistema*), and he said, "Disconnecting from the system *comes from* disconnecting the mind. The Cuban people have always had the saying to disconnect the mind, or just to disconnect. We have always had struggle and we are a happy people, so how do you think we stay happy? We disconnect our minds from our struggles. It's always been, maybe it comes from Africa or something, what do I know."

28 According to Nichter 2010 who paraphrases Raymond Williams (1977), "we always live with ideologies of the past, the present and those emerging on the horizon of a possible future. We likewise live with idioms of distress from the past, which may take on new or hybrid forms and maintain or fade in importance; idioms of the present, associated with the concerns of contemporary life and responses to social change; and emerging idioms. The latter may emerge from several places" (404).

3

Eating in Cuban: The Cuban *Cena* in Literature and Film

Susannah Rodríguez Drissi

> *¡Basta! Os recuerdo el postre. Para eso*
> *no más que el blanco queso,*
> *el blanco queso que el montuno alaba,*
> *en pareja con cascos de guayaba.*
> *Y al final, buen remate a tanto diente,*
> *una taza pequeña*
> *de café carretero y bien caliente.*[1]
>   Nicolás Guillén (from *"Epístola"*)

In August 2006 I returned to Cuba after a 25-year exile in the United States. At the time, I was a first-year graduate student at the University of California, Los Angeles, working on a summer research project on the late Cuban writer and ethnographer Lydia Cabrera. I hoped to see family and childhood friends and, most of all, I hoped to become Cuban again – whatever that meant, under the significant transformations that had taken place since I left the island in the summer of 1981.

With very little time to pull out a notebook or a camera, as I arrived in the José Martí airport, in Havana, I struggled to make a few mental notes, some first impressions that I knew would be lost, were I to neglect the moment of my first return to native soil. Adding to my initial disorientation and the emotional impact of the trip was the visual manifestation that the island and its people had moved on without me – I recognized no one and no one recognized me. I was struck by the transnational feel of the airport itself, unexpected, surely, given the fact that my childhood memories had left me very little to work with, in terms of spatial detail, but also because of the myth of insularity surrounding the island in the last four decades. There were signs in various languages everywhere and the organization of the spaces themselves was decidedly open, even inviting.

With my one-year-old daughter in tow, I looked everywhere for a food venue. I could have used a coffee and she could have used something to entertain her palate and her attention, while we waited for someone to help us with the luggage. Although I certainly didn't expect a *Starbucks*, I was hoping for a *cortadito*. Isn't that an essential of Cuban hospitality? Where were the *pastelitos*? The *masareales*? What happened to the food? Looking for my family in the crowd, I noticed people in the waiting area on their cellphones – they had cellphones! But their faces looked tired: their eyes squinted to overcome the semidarkness of the space, searching for a family resemblance beyond the rope that separated those who arrived from those who were waiting. The space was certainly transnational, but those who waited behind the rope spoke of a Cuban reality beyond my now Cuban-American comprehension.

I had arrived right on time for the carnivals in Havana and on the day of Castro's birthday, but the carnivals had been canceled at the last minute, due to the head of state's ailing health. I had looked forward to the carnivals. They meant, at least as I remember them, crime and drunkards, but also *churros* and *cucuruchos de maní, guarapo* and *serpentina* on the streets – what happened to the carnivals? To Cuban feasting? My understanding of the cultural, political, and economic landscape had certainly grown over the years, but I expected some things to have stayed the same and one of those things had to do with food, with Cuban food and the experience of eating Cuban in Cuba – that is, eating in Cuba the choice menu items found at most Cuban restaurants outside of Cuba: *boliche, lechón azado*, etc; *congrí* and side of fried plantains. As the adage says, "you are what you eat," and I was hoping to be Cuban. I did not want or expect *Pollo Tropical* or *Porto's Bakery*, enclaves of exile eating, at the airport, but somehow the lack of what I considered to be a simple gesture of Cuban warmth upset me. Would my relatives have food waiting for me? And would that food constitute an event worthy of being called a Cuban *cena*? After all, Cubans in Cuba may eat meals everyday (with more or less difficulty and nutritional value), but a *cena* may only be enjoyed on very special occasions. The distinction between the run-of-the-mill plate of Cuban food served and consumed on any given day in Cuba and a Cuban *cena* has to do with four main ingredients: the length of preparation, the abundance of the food prepared, its symbolic value and, last but not least, its transformation into an event. As we neared the entrance to town, I secretly hoped that my family had considered all four.

I arrived in Bauta,[2] my birth town at around 8:30pm; and, as if guessing my thoughts, I was told several times on the ride home that there was a Cuban *cena* waiting for me. What that meant, I knew very well: an overabundance of everything, a full table, many, many dishes, and just as many desserts. It meant eating to the point of gluttony. Certainly, my experience at Cuban parties and *Noche Buenas* at home (and in exile), and my readings in Cuban literature and film underlay my expectations and shaped the significance of the *cena* I was getting ready to eat. My Cuban cultural experience tells me that chary, abstemious eating is an insurance of frailty, ugliness, and even loss of love. To eat Cuban means to devour the food first with the eyes, to look beyond one's plate and fill the plate several times, to satiate both the palate and the appetites of the heart, to overflow the edges of one's senses, to bridge geographical distances. What, then, does it mean to eat a Cuban *cena*?

Since the turn of the millennium, the vast scholarship on the topic of food in film and literature provides some measure of evidence of the ways in which food has become a powerful lens of analysis of both the literary text and the moving picture. For Cuban literature and film, this lens provides fascinating insights into the study of Cuban identity. The role of the Cuban *cena* in literature and film, then, suggests new ways of re-examining the interconnection between cultural practices, national identity, and changing economic conditions. As one of the first examples of the Cuban *cena* in literature, in *Espejo de paciencia* (1608), the first epic poem in Cuban letters, Cuban writer Silvestre de Balboa Troya y Quesada (1563-1649) describes the grand and ample menu that the island offers to Juan de las Cabezas Altamirano, Bishop of Cuba, upon his return from imprisonment. It is there, among the arboreal richness of Puerto Príncipe (now Camagüey) that the moment of return, the "event" that constitutes the moment of return, is celebrated with food – why should my return from the captivity of exile be greeted by anything different?

Troya y Quesada's stanzas read as follow:

Joyful, from the nearby hills,
They come to greet him
All the semicapros of the court,
Satyrs, fauns, and sylvan,
… Offering, with rites and graces
Guanábanas, gegiras and *caimitos.*
… loaded with *mehí*, tobacco,
Mameys, tunas, avocados and pineapples
Mamones and tomatoes and plantains.

The most beautiful and lovely nymphs
Descended in flowing skirts,
With the fruits of *siguapas* and *macaguas*
And many fragrant *pitajayas*;
With fruits of pungent *bijirí* and *jaguas* …
Shrimp, *biajacas* and *guabinas*³… (Canto I, 12-13;
My translation and emphasis)⁴

A "*cena,*" in terms of its symbolic value, Silvestre de Balboa Troya y Quesada's description constitutes an early moment in Cuban letters when the island itself becomes a dining experience, a pageant of gastronomic delights offered to one, in the name of celebration. The *cena*, as the stanzas illustrate, ritualizes, organizes for a chosen few the natural bounty of the island, articulating at this early stage what in the nineteenth century will become the initial markings of a national consciousness in culinary terms. As such, in Cirilo Villaverde's *Cecilia Valdés o La Loma del Ángel*, we find a description of a typical meal served for the Cuban upper classes of the period and given the national importance of the Cuban *cena*:

> The abundance of the viandas matched the variety of the dishes. In addition to the fried beef and pork and two kinds of stew, there was minced veal served in a cassava flour pie, roast chicken, gleaming with butter and seasoned with garlic cloves, fried eggs nearly drowned in salsa [sic tomato sauce), boiled rice, ripe plantain, also fried, in long sweet slices, and a salad of watercress and lettuce.⁵ (Villaverde 2005: 95)

The "abundance" of the food items, in Villaverde's passage, stands for the bounty offered by the island paradise that, in José Martí's own words, in El Diario de campaña, de Cabo Haitiano a Dos Ríos, becomes also a land of (milk and) honey: "Succulent honey in its comb – And throughout the day, what light, what breeze, how puffed-up the chest, how light the sorrowful body! I look beyond the ranch and see, just behind its highest crest, a star and palm tree" (Martí 2007: 81) (My translation).⁶ This brief but telling portrayal of the island's culinary possibilities not only blends with the bounty of its climate and vegetation, but also includes a connection between the "sweets" to be consumed and physical and emotional wellbeing. In the case of this "cena," which includes roasted jutía and other such wild findings, the contentment Martí experiences through the act of eating takes place in the company of the national hero's fellows – that is, in the company of others. It is, after all, for another (and to be shared with another) that a Cuban cena is prepared.

While a cup of Cuban coffee or a Cuban cigar may be enjoyed in solitude, by a veranda, or by a window that opens to the sea, a Cuban *cena* aims to be a centripetal force that calls others into celebration, toward ritual and ceremony, and toward participation in the richness of Cuban culture and the Cuban island itself. The *cena*, through a series of measured, coordinated ingredients and food items on a dish closes on itself and becomes an event. It distances itself from the fortuitous, casual meeting of a fried egg and *picadillo* on a plate to call upon the higher spirit of the symbolic. Indeed, while the trajectory of the Cuban *cena* in literature is related to early representations of Cuban identity as a hodgepodge of ingredients in metaphors, such as Cuban ethnographer Fernado Ortiz's *ajiaco*, or even Celia Cruz's vociferous and playful *¡Azúca!*, the Cuban *cena* is something slightly different, if only because its preparation involves careful planning. There is nothing casual about a Cuban *cena*, neither in times of abundance nor in times of lack. It involves ceremony, and even pomp.

In 2006, the *cena* that was waiting for me in Bauta had been an elaborate combination of elements, clandestine operations, and economic resourcefulness and had required months of planning – it was, at least in spirit, a lezamian *cena*; that is, the same *cena* that during the Republic, Doña Augusta serves her guests at her home in Prado, in the pages of José Lezama Lima's *Paradiso*. Following the small talk that engaged her guests, and a first dish of plantain soup, decorated with a handful of *rositas de maíz* (popcorn), Doña Augusta presents the second entrée, a seafood soufflé, also ornamented, this time with pairs of *langostino* tails:

> The second course appeared, a fluffy soufflé of shellfish, the surface adorned by pairs of prawns forming a circle, their claws spread over the steam coming from the clump like white coral. A paste of giant shrimp caught by our fishermen, who innocently thought that the island's coral platform was entirely crusted with broods of shrimp ... Part of the soufflé also was a fish called emperor, which Dona Augusta had used only because she was tired of porgy; its meat had been extracted first in circles and then by strands; lobsters, showing the livid surprise with which their shells had caught the lantern's inquisition burning their bulging-out eyes. ...
>
> The chill of November, cut by gusts of north wind which ruffled the tops of the Prado poplars, justified the arrival of the glistening turkey, its harsh

extremities softened by butter, its breast capable of attracting the appetite of the whole family and sheltering it as in an ark of the covenant.

And later,

> At the end of the dinner, Doña Augusta tried to be roguish with the dessert. In champagne glasses she served a delicious frozen cream. After the family had paid its most humble homage to the surprise dessert, Doña Augusta gave the recipe. "Simple things," she said, "things that we make in Cuban cooking, the easiest kind of pastry, which the palate immediately declares incomparable. Preserve of grated coconut, another preserve of chopped pineapple, mixed together with a can of condensed milk, and then the good fairy comes, little old Marie Brizzard, to sprinkle the aromatic cream with an-isette. Into the refrigerator, serve it up quite cold. Then we set about tasting it, receiving the praise of the other diners as they insist on an encore, just as when we hear one of Lully's pavanes." (Lima 2000: 181-183)[7]

It is only after the experience of dining at Doña Augusta's has taken place (and after coffee and cigars) that the guests venture out into the night breeze, "into the cold in front of the house, where they could see the waves rolling in broad barrels,"[8] breaking against the wall of the *malecón*. The accumulation of dishes and general culinary presentation in this Cuban *cena*, which includes plates, pristine tablecloths, glasses, decorative objects, etc., contributes to a sensation of rootedness and to the creation of a closed or enclosed stage – a historical, sociocultural and affective space. Far from creating a feeling of claustrophobia or captivity, the *cena* constitutes a space of belonging and elite membership into "*lo cubano.*"

While the cultural meaning and symbolic value of the *cena* enjoy a certain level of stability, in spite of political, economic, and geographical changes, as the following examples from film and literature will show, in the decades that follow the Cuban Revolution of 1959, the menu of the Cuban *cena* undergoes some significant transformations. The *lezamian cena*, Lezama Lima's contribu-tion to the Cuban literary imagination – exemplary in Cuban literature, in terms of its culinary affections – re-emerges numerous times in the next few decades, its menu adjusted to fit the politics and economy of the times, as a scene in Tomás Gutiérrez Alea's *Fresa y chocolate* (1994) reveals. In the scene that follows, Diego, the openly antirevolutionary homosexual, invites David, a young writer and communist who visits Diego's *guarida* for various reasons, one being a much needed lesson in good writing, to eat *chez lui*. The invitation to a lezamian *cena* is not without worry, as the items needed for the meal are not only expensive but difficult to find in Cuba in the late 1970s. Following is the passage when Diego asks his neighbor Nancy to make a few purchases:

> "But Dieguito, a lezamian lunch? That's gonna be about a hundred dollars!"
> "And a small fortune, may I add." (Alea and Tabío 1995)[9]

Later, Diego explains to David the importance of the meal they are about to share:

> You are partaking of the lunch that Doña Augusta offers her guests in the pages of *Paradiso*, Chapter Seven. After this, you will be able to say that you have eaten like a real Cuban, entering forever into covenant with those who adore the Master, lacking only the knowledge of his work. Next, we ate roast turkey, chilled cream (also of lezamian origin), the recipe of which he gave me, so that I, in turn, would pass it on to my mother. Now, Baldovina would have brought the fruit bowl but, missing Baldovina, I'll bring it. You must forgive me the apples and pears, which I have substituted for mangoes and guavas, a nice addition to the mandarins and grapes. Later, we'll drink coffee outside on the balcony, while I read you poems by Zenea, the vilified. (Alea and Tabío 1995)[10]

This *cena* is, in many ways, a Last Supper, the significance of which Gutiérrez Alea had already played with in his early film, *La última cena* (1976). The *cena*, which, in the context of the film becomes an *almuerzo* or lunch, represents a goodbye to the island and to friends, for Diego knows that he will be leaving Cuba soon. Most importantly, it constitutes David's rite of initiation, a transmission of cultural knowledge that introduces him to Cuban culture and allows him to participate in both its culinary and literary pleasures. Furthermore, Diego equips David with its recipe, a text of sorts that may be read, interpreted and reinterpreted, depending on a given set of social, economic, and symbolic frameworks. From this point forward, David will need to determine with whom he will share the *cena*, and what he will be able – given the country's present and future economy – to include in the menu. The meaning he ascribes to it, regardless of pomp and circumstance, will always have to do with Cuban identity, the instability of which the *cena*, as ritual, proposes to "ground" or stabilize.

Indeed, the lezamian *cena* Diego prepares is not without substitutions, as the indigenous *mangoes* and *guavas* take the place of the foreign, such as apples and pears. As such, literary tradition, cultural innovation, and economic limitations matter equally; the lezamian recipe is modified, if only slightly, in the name of Cuban adaptability, as Michel de Certeau and Luce Giard suggest, in "The Nourishing Arts," "to serve the needs of the hour, to furnish the joy of the moment, and to suit the circumstances" (Certeau and Giard 1997: 69).

The Cuban *cena*, as the passage reveals – much like other habits and activities Cubans may engage and find enjoyment in, such as dancing and playing domino – is deeply rooted in the fabric of relationships to others and to the cultural matrix of the island. Diego's *cena*, in this sense, has more to do with cultural and emotional nourishment than with dietary needs and, as such, its nutritional value is never considered. Nowhere is there mention of the fat, sugar, and carbohydrate components of the meal. In fact, if we consider that a traditional Cuban *congrí* (rice and beans) includes a large amount of pork lard, in order to give the final product a satin finish, then we quickly conclude that the Cuban *cena* pays very little attention to the possible negative effects its menu could have on the body and its proper functioning. Furthermore, the temporal (event vs. everyday) and performative quality (ritual and artifice) of the *cena* suggests the time of carnival; that is, the breaking of fast. However, while in David's *cena*, we also find the suggestion of the carnivalesque in the collapse of the differing political affiliations and sexual orientations represented (the gay antirevolutionary sits at the same table with the communist and heterosexual David), the Cuban *cena*, as planned and prepared by Diego, is an expression of cultural identity that aims to replace or at least compete with David's political membership.

But, what exactly does this mean? A careful look at the literary representation of the Cuban *cena* in the decades to follow may suggest the answer.

In Zoe Valdes's first novel, *Yocandra in the Paradise of Nada* (*La nada cotidiana*), published in 1995, eating like a Cuban in Cuba extends the substitutions of apples and pears we find in Diego's lezamian *almuerzo* to also include the substitution of the entire culinary components of the Cuban *cena*, the adaptability of which does very little to upturn its symbolic and cultural value. Eating a Cuban *cena* like a Cuban in Cuba in Valdes's novel has a great deal to do with the economic and political situation of the island. The menu Yocandra shares with her on-and-off partner, The Nihilist, is one made up of pizza laden with a cheese substitute, "smeared with some reddish liquid that bears no relationship whatever to tomato sauce," and a bottle of Beaujolais

Nouveau, "Real French wine" (Valdés 1997: 129)! The Nihilist explains:

> I had to wait so long in line for pizza at the Piragua that my clothes had time
> to dry. The pizza delivery was late today. Incredibly late. You wouldn't believe
> how many people were in line. If I had waited my turn, I would never have
> got any pizza. So I bought it from a scalper. You know what I had to pay?
> A hundred and twenty pesos! Sixty for each pizza! That's only 30 pesos less
> than my entire monthly salary. And the cheese isn't real cheese, it's some
> Chinese substitute. The people who work at the place steal all the real cheese.
> (Valdés 1997: 128)[11]

Dumbfounded, Yocandra watches her lover take two pieces of bread out from
his knapsack, "each more or less round, each a different size, both reeking of
rancid cheese" (Valdés 1997: 128)[12] – that is, cheese substitute. She puts them
immediately into the oven, in case the gas suddenly goes off, takes them out
a minute later and puts them in two frying pans. Although class and historical
differences may account for some of the substitutions and variations (settings,
characters, etc.) between the original lezamian *cena* and Yocandra's *cena*, its
cultural and symbolic value remains essentially the same: the Cuban *cena* continues to be a group ritual, a unifying structure in social, spatial, and affective
terms. The *cena* as "event" is underscored by the simple fact that to drink real
French wine and eat pizza (even if laden with cheese substitute) in Cuba in the
1990s constitutes an event in itself, the symbolic significance of which may be
suggested, among other things, by the accompanying sexual feasting that ensues. "Abundance" in this sense is a creative and thus imaginative component
without which the *cena* would not be possible.

It is of significant importance to note, however, that the event, while historically and symbolically grounded in local culture, aims to take place beyond
the local, as the lovers imagine that they are in Europe and not Cuba. To eat a
Cuban *cena* in Valdes's text, then, is to transcend the local economic reality of
the island, to break insularity and move beyond Cuba's geographical borders.
Yocandra rejects the Chinese substitution of the cheese, not because it is Chinese, but because it is a substitution of what she considers to be real cheese.

Much like the Cuban *cenas* seen thus far, then, Yocandra's *cena* with The
Nihilist is marked by ceremony, as Yocandra lights candles and puts on
a record of Gregorian chants to stage the event, setting the table with an
embroidered white tablecloth and napkins, "a gift from the Gusana's [a friend
exiled in Spain] grandmother to her parents on their forty-something wedding

anniversary." (Valdés 1997: 129)[13] In this case, the history of the tablecloth Yocandra uses to stage the scene/event lends new textures of significance to the *cena*, imparting a richer field of meaning that, unlike Doña Augusta's white tablecloth in Lezama's *Paradiso*, also includes a clearly political dimension: as Yocandra confesses, "[e]ating pizza in Havana is like dining at the Tour d'Argent in Paris. Not only because you have to make a reservation to eat in a ramshackle pizza parlor, but you must be high up in the Militant Union pecking order" (Valdés 1997: 129).[14] As for the tablecloth, it is an heirloom of exile, the politics of which the numerous migratory waves of Cubans to all corners of the world corroborate.

In this way, the *cena* becomes a unifying drama, the drama that ensues from the effort of turning the everyday into an event at a time when Cuba's economy is, at best, dwindling. The burned pizza with substitute cheese and what attempts to pass for tomato sauce, accompanied by French wine, candles, music and a great deal of imagination constitute a Cuban *cena* worthy of a moment that also includes a sexual act:

> his hands take refuge on my loins. There's the smell of something burning. We eat the half-burned pizzas. We have won the battle of the beans. (Valdés 1997: 129)[15]

The physical encounter between Yocandra and The Nihilist, complemented by a description of her own physical attributes as "loins," underscores the significance of the *cena* as an act of human bonding. The potential encounter between Diego and David in Gutiérrez Alea's film becomes a reality and yet another component of the Cuban *cena* in Valdés – in other words, in *Yocandra and the Paradise of Nada*, the body is also consumed and the sexual encounter that ensues participates in the necessary confusion between fantasy and reality that the *cena* ultimately establishes:

> The Nihilist's prick – ¡*Ayé!* Blessed Lazarus, my Babalú – is the eighth wonder of the world. It could make him the head of one of the greatest fortunes of our century. Being endowed with a prick like that is like having millions in the Swiss bank. ... And when your hands make contact with the base of the member, your mouth waters and your lips foam. You just can't help it.
>
>   It's smooth. It measures six and a half inches at rest, twice when standing at attention. ... The shaft is solid, unperturbed by any unexpected movements of the earth and bolstered by the centuries, a true pillar of the Panthe-

on. … It exudes an odor of skin washed with Mon Savon, that French soap concocted of ancient perfumes: essence of patchouli, jasmine, roses, goat's milk. Milk, my beloved's milk! If only our daily rations were at the same level! … Milk, milk of my heart! This man's milk seems to spring forth as from a young Holstein, falling like manna from heaven, extraterrestrial sperm that gives you a mouthful of stars, shining brightly, interplanetary, transmitted by satellite. A cocktail of spermatozoa. (Valdés 1997: 125-126)[16]

Even the performed sexual act takes Yocandra far beyond the island, as Diego admits in Gutiérrez Aleas's film, "on the wings of imagination." Her lover's Milk in the dilapidated apartment in Havana does for Yocandra what wild honey does for Martí in the Cuban heartland: it assuages – in very wide terms – the appetite. Winning the "battle of the beans," then, means to overcome the prosaic, the cruel economic reality of Cuba in the Special Period in Times of Peace, the euphemism used to describe the economic conditions on the island following the fall of the Berlin Wall and the collapse of the Soviet Bloc. Here, too, there is very little concern with the nutritional value of the food consumed. Beans, one of the only staples of traditional Cuban cuisine available in the 1990s, became a nutritional necessity, as protein in any other form was scarce and, at times, almost impossible to find. Far from possessing the nutritional value of beans, the pizza and real French wine that establish the menu, along with the sexual accoutrements of the moment are enough to stage a Cuban *cena*, and the event – in cultural and affective terms – that it signifies.

The imaginative element of eating Cuban is taken one step further in the work of Abilio Estévez. Published in 2002, *Los palacios distantes*, Estévez's second novel, takes place in the ruined topography of Havana City, and tells the story of Victorio, a 40-something homosexual who is evicted from his apartment days before the building collapses. As a response to the notice of eviction, Victorio burns almost all of his possessions and abandons his job. Deprived of the most basic necessities, but armed with the knowledge that his own private palace awaits him in some unknown Elsewhere, Victorio wanders through the city in ruins, until he meets Salma, a beautiful young "*jinetera*,"[17] or Cuban sex worker. Sometime later, Victorio encounters an old clown, whom he decides to follow. Homeless and without a job, Victorio is finally overcome by hunger and exhaustion and faints, to later wake up in the presence of the old clown, Don Fuco, and in the ruins of an old theater, the *Pequeño Liceo*. Don Fuco, who lives in the ruins of the theater, spends his time making people laugh, showing up at hospitals and hospices, and appearing and disappearing in cemeteries and wakes, remedying the disillusion and hopelessness of Cubans in the year 2000 with artifice and hope.

Together, Salma and Victorio join Don Fuco in the ruins of the theatre where the clown trains them in the art of buffoonery. The old theater, a repository of Cuban culture, houses, among its many treasures, the wardrobe of Benny Moré, Celia Cruz, Alicia Alonso, Rita Montaner, and Barbarito Diez, to name only a few – as well as guitars, a piano, canvases, and the manuscripts of famous writers (Estévez 2002b: 135). It is precisely in the theater where the Cuban *cena* takes place:

> "Come, follow me, I'm going to show you something."
> Salma takes Victorio by the arm. She leads him to the stage.
> Where a large dining table is set for a sumptuous meal. The Bruges lace tablecloth is splendid. Candelabras with spiraling candles and trembling flames. Limoges china. Silver dinnerware. Diaphanous crystal glasses from Bohemia. (Estévez 2002b: 220)

Up until this moment, the *cena* that the friends are about to enjoy is almost identical to the Cuban *cenas* seen earlier: the table, linens and silverware, along with lit candles and heirlooms set the stage for an extraordinary experience. The events that follow, however, alert us to a change in the material and economic conditions surrounding the encounter:

> Don Fuco welcomes them. Don Fuco spreads his arms. In silence, Salma and Victorio stand behind the chairs that have been assigned to them, waiting for the old clown to raise his empty cup and tells them what he wants to toast. The clown merely laughs. His forthright laughter seems to be sufficient. Ceremoniously, without a single word, they all lift up their empty glasses. They sit down. They begin to laugh. No one pours white wine, or red wine, or even water into the Bohemia glasses. No one carries in any platters of food. They sit there at a sumptuously set table on which there is no food nor drink. They see the bottoms of their soup bowls, where plump, rosy nymphs, pursued by lusty satyrs, run cheerfully among absurd trees. They can't stop laughing. Sometimes they seem to hear applause. In the middle of laughing, Salma manages to say that the applause is really the beating of all the doves' and bats' wings. Between guffaws, Victorio explains, "No, no, don't you realize? It's the rain." More calmly, Don Fuco asserts, "Don't

doubt it: it is applause, old applause imprisoned between these walls." (Estévez 2002b: 220)[18]

In much the same way as the *cena* (or *almuerzo*) in Gutiérrez Alea's *Strawberry and Chocolate*, Estévez's *cena* constitutes a rite of initiation, as Salma and Victorio become participants in the means and ways of Cuban culture, as represented in the ruins of *El Pequeño Liceo*. Here, however, there is an emphasis on imagination and cultural memory, as the ordinary act of sitting at a table in Cuba at the turn of the twenty-first century becomes an extraordinary event where abundance depends entirely on unspoken individual and group recollections of plenty. The nymphs that adorn the plates recall the playful sprites who, in Silvestre de Balboa Troya y Quesada's *Espejo de paciencia*, (whose stanzas introduce the present discussion), come bearing the natural gifts that the island may offer to one recently freed from captivity. The applause tells us, in no uncertain terms, that a performance of culture has taken place.

Likewise, my own *cena* with family members in Bauta – which also included a large portion of laughter and *choteo*[19] – allowed me to say, much like Alea's David, that I had eaten "*como un[a] real cubano[a]*" (like a real Cuban) and praise was implied; in my case, however, this meant that I had eaten "Cuban" in Cuba. It seems that, for me, cultural authenticity and membership into *lo cubano* are inextricably linked to the relationship between food and geography, so that being Cuban has to do with eating Cuban like Cubans in Cuba. At first glance, however, the very notion of the Cuban *cena* does not permit this to happen. The *cena*, as seen thus far, transcends, even avoids the quotidian. The irony of my connection lies in the fact that had I (in 2006) been allowed to eat like a Cuban in Cuba, I wouldn't have been served the same menu.

"Home" for the summer, after 25 years I, too, had enjoyed a Cuban *cena* worthy of applause, not because the menu was extraordinary (*tamales, congrí,* roasted pork, *casquitos de guayaba, dulce 'e coco* and cream cheese), but because of the extraordinary efforts that had gone into preparing it, the ritual of return it constituted for me, and because of the necessary reconstruction (real and imagined) of tastes, gestures, emotions that make up what eating Cuban in Cuba (and abroad) means to me.

References

Alea, T.G. and Tabío, J.C. (1995). *Fresa y chocolate*.
Cabezas, A. (2009) *Economies of Desire: Tourism and Sex in Cuba and the*

Dominican Republic. Philadelphia, PA: Temple University Press.

De Certeau, M. and Giard, L. (1997) 'The Nourishing Arts', in C. Counihan and P. Van Esterik (eds.) *Food and Culture. A Reader*, 2nd edn, New York and London: Routledge.

Estévez, A. (2002a) *Los palacios distantes*, Barcelona: Tusquets.

— (2002b) *Distant Palaces*, trans. David Frye, New York: Arcade Publishing.

Lima, J.L. (1968) *Paradiso*, Mexico: Biblioteca Era.

— (2000) *Paradiso*, trans. Gregory Rabassa, Illinois State University: Dalkey Archive Press.

Mañach y Robato, J. (1928) *La crisis de la alta cultural en Cuba: Indagación del choteo*, 1st edn, La Habana: Revista de Avance.

Martí, J. (2007) *Diario de campaña*, Barcelona: Linkgua Ediciones S.L.

Muñoz, J.E. (1999) *Disidentifications: Queers of Color and the Perfomance of Politics*, Minneapolis, MN: University of Minnesota Press.

Troya y Quesada, S. (2011) *Espejo de paciencia*, Barcelona: Red Ediciones, S.L.

Valdés, Z. (1995) *La nada cotidiana*, Barcelona: Salamandra.

— (1997) *Yocandra in the Paradise of Nada—A Story of Cuba*, trans. Sabina Cienfuegos, New York: Arcade Publishing.

Villaverde, C. (2005) *Cecilia Valdés or El Angel Hill*, trans. Helen Lane, New York: Oxford University Press.

— (2011) *Cecilia Valdés o la loma del ángel*, Barcelona: Red Ediciones, S.L.

Endnotes

1   *"Enough! May I remind you of dessert / For which white cheese may only do, / The pale cheese so esteemed up on our hills / Paired with a side of guava shells / And, to close it off, after much chewing, /A small cup of boiled coffee, hot and brewing."* (My translation)

2   It may be of interest to recall that Bauta is that little town on the outskirts of Havana where the members of *Orígenes* (1944-1954), led by José Lezama Lima, often gathered for dinner and intellectual feasting at the house of Father Ángel Gaztelu.

3   River fish; endemic species of Cuba.

4   *"Sálenle a recibir con regocijo / De aquellos montes por allí cercanos / Todos los semicapros del cortijo, / Los sátiros, los faunos y silvanos / [...] Le ofrecen frutas con graciosos ritos, / Guanábanas, gegiras y caimitos. / [...] Vienen cargadas de mehí y tabaco, / Mameyes, pinas, tunas y aguacates, / Plátanos y mamones y tomates. / Bajaron de los árboles en na- guas / Las bellas hamadriades hermosas / Con frutas de siguapas y macaguas / Y muchas*

pitajayas olorosas; / De bijirí cargadas y de jaguas / [...] Camarones, biajacas y guabinas [...]' (Troya y Quesada 2011: 12-13).

5   *La abundancia de las viandas corría pareja con la variedad de los platos. Además de la carne de vaca y de puerco frita, guisada y estofada, había picadillo de ternera servido en una torta de casabe mojado, pollo asado relumbrante con la manteca y los ajos, huevos fritos casi anegados en una salsa de tomates, arroz cocido, plátano maduro también frito, en luengas y melosas tajadas, y ensalada de berros y de lechuga.* (Villaverde 2011: 100)

6   *"Rica miel, en panal. – Y en todo el día, ¡qué luz, qué aire, qué lleno el pecho, qué ligero el cuerpo angustiado! Miro del rancho afuera, y veo, en lo alto de la cresta atrás, una palma y una estrella"* (Martí 2007: 81).

7   *Hizo su entrada el segundo plato en un pulverizado* soufflé *de mariscos, ornado en la superficie por una cuadrilla de langostinos, dispuestos en coro, unidos por parejas, distribuyendo sus pinzas el humo brotante de la masa apretada como un coral blanco. Una pasta de camarones gigantomas, aportados por nuestros pescadores, que creían con ingenuidad que toda la plataforma coralina de la isla estaba incrustada por camadas de camarones[] Formaba parte también del* soufflé, *el pescado llamado emperador, que dona Augusta sólo empleaba en el cansancio del pargo, cuya masa se había extraído primero por círculos y después por hebras; langostas que mostraban el asombro cárdeno conque sus carapachos habían recibido la interrogación de la linterna al quemarles los ojos saltones. [] El friecito de noviembre cortado por ráfagazos norteños que hacían sonar la copa de los álamos del prado, justificaba la llegada del pavón sobredorado, suavizada por la mantequilla las asperezas de sus extremidades, pero con una pechuga capaz de ceñir todo el apetito de la familia y guardarlo abrigado como en un arca de la alianza [] Al final de la comida, doña Augusta quiso mostrar una travesura en el postre. Presentó en las copas de champagne la más deliciosa crema helada [] Son las cosas sencillas - dijo -, que podemos hacer en la cocina cubana, la repostería más fácil, y que enseguida el paladar declara incomparable. Un coco rayado el conserva, más otra conserva de piña rayada, unidas a la mitad de otra lata de leche condensada, y llega entonces helada, es decir, la viejita Mariebrizard, para rociar con su anisete la crema olorosa. Al refrigerador, se sirve cuando está bien fría. Luego la vamos saboreando, recibiendo los elogios de los otros comensales que piden con insistencia el* bis, *como cuando oímos alguna pavana de Lully.* (Lima 1968: 196-198)

8   *"al frío del portal, desde donde se divisan las olas [...]"* (Lima 1968: 184).

9   *"Pero Dieguito, ¿un almuerzo lezamiano? ¡Eso es como cien dólares!"*
      "Una pequeña fortuna en Cuba, agrego."

10   *"Estás asistiendo al almuerzo que ofrece doña Augusta en las páginas de* Paradiso, *capítulo séptimo. Después de esto podrás decir que has comido como un real cubano, y entras, para siempre, en la cofradía de los adoradores del Maestro, faltándote, tan sólo, el conocimiento de su obra. A continuación comimos pavo asado, seguido de crema helada también lezamiana,*

de la que me ofreció la receta para que yo a mi vez la transmitiera a mi madre. Ahora Baldovina tendría que traer el frutero, pero a falta suya iré por él. Me disculparás las manzanas y las peras, que he sustituido por mangos y guayabas, lo que no está del todo mal al lado de mandarinas y uvas. Después nos queda el café, que tomaremos en el balcón mientras te recito poemas de Zenea, el vilipendiado [ ]"(Alea and Tabío 1995).

11   *"La ropa se me escurrió encima, esperando la cola de la Piragua. Las pizzas llegaron tardísimo, había una concentración de gente… si me quedaba en la cola no alcanzaba pizza… ¿Sabes cuánto pagué a un colero por dos pizzas? Ciento veinte pesos, sesenta cada una. Menos treinta pesos, es mi salario de un mes… Y el queso no es queso, es preservativo chino derretido. Los empleados se fachan el queso bueno"* (Valdés 1995: 144-145).

12   *"[…] mal redondeados, desiguales en tamaño, apestosos a queso […]"* (Valdés 1995: 145).

13   *"[…] regalo de boda de la bisabuela de la Gusana a los padres de ésta, allá por el cuarenta y pico"* (Valdés 1995: 146).

14   *"Comer pizza en La Habana en estos tiempos equivale a cenar en el famoso restorán parisino* La tour d'argent. *Para comer en una pizzería andrajosa hay que reservar un turno y ser obrero destacado en el Sindicato"* (Valdés 1995: 145).

15   *"[…] las manos caen cruzadas sobre los hoyuelos traseros de mis caderas. El pan comienza a oler a quemado y servimos las pizzas a punto de devenir tizones. Somos unos privilegiados, le ganamos la batalla al chícharo."* (Valdés 1995: 145)

16   *"La pinga, ¡ay, San Lázaro bendito, mi Babalú Ayé! El toletón del Nihilista es la octava maravilla del mundo. ¡Y cuida'o no ocupe el primer escaño en el escalafón de las fortunas de este siglo! Porque portar un rabo como ése es como poseer una cuenta de millones y millones de dólares en un banco suizo. Debo señalar, antes de que lo olvide, que junto al obligo tiene un lunar negro y redondo. Y desde allí le emerge de los poros la sedosa pendejera que es un sueño acariciarla. Cuando la mano tropieza con la raíz del miembro — nada que ver con un miembro del cedeerre — no puedes evitarlo, la boca se te hace agua, las comisuras espumean. Es liso. Mide catorce centímetros sin erección, el doble erecto. […] El centro es sólido, a prueba de derrumbes, apuntala'o desde siglos inverosímiles a.n.e., semejante columna del Partenón. […] Exhala un perfume a piel lavada con* Mon savon, *ese jabón francés a base de extractos de fórmulas de perfumes antiquísimos, pachulí, jazmín, rosas, leche de cabra. ¡Leche, leche mía! ¡Si la cuota de dieta fuera como ésa! ¡Leche, leche de mi corazón! La savia de este hombre es como cuando hordeñan a una Holstein jovenzuela, y el chorro cae en vasinilla igualito al maná celestial. Ése es precisamente el sabor de la esperma de este extraterrestre, un buche estrellado, luminoso, interplanetario, vía satélite. Un ponche repleto de una turba de saludables, deportivos, y preñadores espermatozoides"* (Valdés 1995: 142-143).

17   The word *jinetera*, which comes from the Spanish word *jinete* (jockey), is a Cuban slang word used to refer to Cuban prostitutes. The word first entered

Cuban lexicon in the 1990s and is associated with the economic crisis following the fall of the Berlin Wall and the end of the Soviet Bloc. For a more complete definition of the term and its racial implications, please see Amalia L. Cabezas's *Economies of Desire: Sex and Tourism in Cuba and the Dominican Republic* (2009).

18   "*Don Fuco los recibe. Don Fuco abre los brazos. En silencio, Salma y Victorio se detienen frente a las sillas que les han sido asignadas, a la espera de que el anciano payaso se levante la copa vacía y diga por qué desea brindar. El payaso sólo ríe. La risa franca parece suficiente. Ceremoniosos, sin una palabra, elevan las vacías copas al frente. Se sientan. Comienzan a reír. Nadie pone vino, blanco o tinto, ni siquiera agua, en las copas de Bohemia. Nadie trae bandeja alguna con comida. Están allí, frente a la suntuosa mesa en la que no hay licores ni alimentos. Ríen. No cruzan palabra. La risa no les permite hablar. Miran el fondo de los platos soperos, donde ninfas rollizas y sonrosadas, perseguidas por sátiros lascivos, corren con alegría por entre los árboles absurdos. No pueden parar de reír. A veces parece como si escucharan aplausos. En medio de tanta risa, Salma encuentra un momento para decir que los aplausos son en realidad el aletear de tantas palomas y murciélagos. Entre carcajadas, Victorio aclara No, no, ¿no se dan cuenta?, es la lluvia. Más calmado, Don Fuco afirma No lo duden, son aplausos, aplausos antiguos apresados entre estas paredes*" (Estévez 2002a: 231).

19   *Choteo* means joking or jesting. It is a Cuban expression, claimed by the Cuban ethnographer Fernando Ortíz (1924) to derive from Afro-Cuban idiomatic speech that variously connotes the act of "tearing, talking, throwing, maligning, spying, and playing," according to the Cuban American writer José Muñoz (1999). Jorge Mañach y Robato's essay *La crisis de la alta cultural en Cuba: Indagación del choteo* (1928), provides an extensive study of its role in the social matrix of Cuban culture.

Section II

Social Accountability and "the Gift": Internationalism and Cuban Medical Diplomacy

II

4

Fifty Years of Cuba's Medical Diplomacy: From Idealism to Pragmatism[1]

Julie M. Feinsilver

Prologue

Twenty years ago, I examined 30 years of Cuban medical diplomacy based on research that I had conducted over the previous decade (Feinsilver 1989a; 1989b). In reviewing the past 50 years of Cuban medical diplomacy for this article, I revisited my earlier writings of 1989; my 1993 book, *Healing the Masses: Cuban Health Politics at Home and Abroad*;[2] a series of articles I have written over the past four years; and new data (Feinsilver 2009a, 2009b, 2008, 2006, 1993). At the conceptual level, the more things changed, the more they have stayed the same.

Clearly, there have been several key changes affecting Cuba over the past two decades. Since 1989, we have witnessed the collapse of the Soviet Union and the web of trade and aid relationships Cuba had with the countries in the old Soviet sphere of influence. The resulting economic crisis for Cuba made the decade of the 1990s, to a large extent, a lost decade. Then, just in the nick of time, Hugo Chávez came to power in Venezuela in 1998, ushering in a new era of preferential trade and aid agreements that provides economic largesse for Cuba. The global financial and economic crisis that began in late 2007 and three devastating hurricanes in 2008 have again thrown Cuba's economy into a tailspin. What stayed the same is the subject of this article: Cuba's commitment to and conduct of medical diplomacy.

Introduction

Medical diplomacy, the collaboration between countries to simultaneously produce health benefits and improve relations, has been a cornerstone of Cuban foreign policy since the outset of the revolution 50 years ago. It has helped Cuba garner symbolic capital - goodwill, influence, and prestige - well beyond what would have been possible for a small, developing country, and it has contributed to making Cuba a player on the world stage. In recent years, medical diplomacy has been instrumental in providing considerable material capital - aid, credit, and trade - to keep the revolution afloat. This analysis examines why and how Cuba has conducted medical diplomacy over the past 50 years, the results of that effort, and the mix of idealism and pragmatism that has characterized this experience. A revolution can be measured by its actions to implement its ideals, something the Cubans have done successfully through medical diplomacy.

The Nature of Cuban Medical Diplomacy

Enabling factors

From the initial days of the revolutionary government, Cuba's leaders espoused free universal health care as a basic human right and responsibility of the state. They soon took this ideological commitment to the extreme and contended that the health of the population was a metaphor for the health of the body politic. This assertion led to the establishment of a national health system that, over time and through trial and error, has evolved into a model lauded by international health experts, including the World Health Organization (WHO). The Cuban health system has produced key health indicators, such as infant mortality rate and life expectancy at birth, comparable to those of the United States, even though there is a vast difference in the resources available to Cuba to achieve them.

At the same time, Cuban health ideology always has had an international dimension. It has considered South-South cooperation to be Cuba's duty as a means of repaying its debt to humanity for support it received from others during the revolution. Therefore, the provision of medical aid to other developing countries has been a key element of Cuba's international relations despite the immediate postrevolutionary flight of nearly half of the island's doctors and the domestic hardship this aid may have caused.

The medical brain drain contributed to the government's decision to reform the health sector, revamp medical education, and vastly increase the number of doctors trained. These factors combined made possible the large-scale commitment to medical diplomacy and lent credibility to Cuba's aid offers. They also demonstrated Cuba's success on the ground in reducing mortality and morbidity rates, which are primary goals of all health care systems. By the mid-1980s, Cuba was producing large numbers of doctors beyond its own health care system needs specifically for its internationalist program. The latest available data from late 2008 indicate that Cuba has one doctor for every 151 inhabitants, a ratio unparalleled anywhere (Oficina Nacional de Estadísticas 2008).

Cuba's initial foray into medical diplomacy

Despite Cuba's own economic difficulties and the exodus of half of its doctors, Cuba began conducting medical diplomacy in 1960 by sending a medical team to Chile to provide disaster relief aid after a major earthquake. Three years later, and with the U.S. embargo in place, Cuba began its first long-term medical diplomacy initiative by sending a group of 56 doctors and other health workers to provide aid in Algeria on a 14-month assignment. Since then, Cuba has provided medical assistance to more than one hundred countries throughout the world both for short-term emergencies and on a long-term basis. Moreover, Cuba has provided free medical education for tens of thousands of foreign students in an effort to contribute to the sustainability of its medical assistance.

Perhaps as a portent of things to come, even during the 1970s and 1980s, Cuba implemented a disproportionately larger civilian aid program - particularly medical diplomacy - than its more developed trade partners: the Soviet Union, the Eastern European countries, and China. Cuban civilian aid workers constituted 19.4% of the total provided by these countries, although Cuba accounted for only 2.5% of the population. This quickly generated considerable symbolic capital for Cuba, which translated into political backing in the UN General Assembly, as well as material benefits in the case of Angola, Iraq, and other countries that could afford to pay fees for professional services rendered, although the charges were considerably less than market rates (Feinsilver 1993: 159-160).

Typology of Cuba's medical diplomacy

The following typology of Cuba's medical diplomacy initiatives facilitates understanding at a glance the depth and breadth of Havana's use of this type of soft power as a major instrument of foreign policy. Cuba's aid can be divided into two major categories: short-term and long-term initiatives, although some short-term initiatives may have long-term effects. Moreover, some subcategories are not mutually exclusive as the particular initiative may be conducted on either a short-term or long-term basis (Feinsilver 1993).

Short-term initiatives can then subdivided into nine categories:

1.   Disaster relief
2.   Epidemic control and epidemiological monitoring
3.   On-the-job training for health care professionals to improve their skills
4.   Direct provision of medical care in Cuba
5.   Health system organizational, administrative, and planning advisory services
6.   Donation of medicines, medical supplies, and equipment
7.   Vaccination and health education campaigns
8.   Program design for human resource development and for the provision of specific medical services
9.   Exchange of research findings and knowledge transfer through the sponsorship of international conferences and the publication of medical journals

Cuba's long-term medical diplomacy initiatives can be categorized in seven areas:

1.   Direct provision of primary health care in the beneficiary country, particularly in areas where local doctors will not work
2.   Staffing of secondary and tertiary care hospitals in beneficiary countries
3.   Establishment of health care facilities (e.g., clinics, diagnostic laboratories, hospitals) in beneficiary countries
4.   Establishment of comprehensive health programs in beneficiary countries
5.   Establishment and/or staffing of medical schools in beneficiary countries and/or in-country community clinic-based medical education combined with distance learning under Cuban supervision in country
6.   Provision of full scholarships to study in Cuba for medical school and allied health professional students
7.   Scientific exchanges

Already by the mid-1970s, Cuba had used all but two of the foregoing instruments. Not developed until much later, the two are establishment of comprehensive health programs in country, an approach first developed in 1998, and in-country community clinic-based medical education combined with distance learning, the Virtual Health University established in 2006. A few examples of some of these initiatives over the past 50 years demonstrate that, although Cuba is a small, developing country, it has been able to conduct a world-class foreign policy from which President Barack Obama recently said the United States could learn (Blanchfield 2009).

Disaster relief

Cuba has been quick to mobilize well-trained disaster relief teams for many of the major disasters in the world. Among its recent activities were specially trained disaster relief medical brigades - 60 doctors - immediately dispatched to Haiti after the January 2010 earthquake to supplement the existing 400-strong medical brigade and more than 500 Haitian graduates of Cuban medical schools who worked with them. Because the Cuban doctors were already working in all ten departments in Haiti and teams of Cuban doctors had worked in country since 1998, they were the first foreigners to respond to the great earthquake. After three weeks, they had assisted over 50,000 people; conducted 3,000 surgeries, 1,500 of which were complex operations; delivered 280 babies; vaccinated 20,000 people against tetanus; established nine rehabilitation wards; and began providing mental health care, particularly for children and youths ('Cuban doctors in Haiti' 2010; '*Cuba considera*' 2010).

After Hurricane Georges devastated Haiti in 1998, Cuba also was the first country to send medical aid to Haiti. After immediate disaster-relief work, Cuba began providing free medical care to the Haitian people on a long-term basis, implementing their model Comprehensive Health Program, and providing full scholarships to Haitian medical students for study in Cuba. In response to Tropical Storm Jeanne in 2004, Cuba sent an additional team of 64 doctors and 12 tons of medical supplies to Haiti. Between 1998 and 2010, 6,094 Cuban medical professionals have worked in Haiti, conducting more than 14 million patient visits, 225,000 surgeries, 100,000 birth deliveries, and saving more than 230,000 lives. In addition, by 2010, with Venezuelan support, Cuba had established five of ten planned comprehensive diagnostic centers, which provide not only a range of diagnostic services but also emergency care. Cuba had also trained 570 Haitian doctors on full scholarships. Finally, since 2004, slightly more than 47,000 Haitians have undergone free eye surgery as part of Operation Miracle (Embassy of Cuba in Nueva Zealand 2010).

Other recent disaster areas to which Cuba deployed its specialized medical brigades are China after the May 2008 earthquake, Indonesia after the May 2007 earthquake, Bolivia after the February 2008 floods, and Peru after the December 2007 earthquake. Cuban medical missions provided assistance as well in post-2004-tsunami Indonesia and post-2005-earthquake Pakistan. In both cases, the Cuban medical teams initially provided disaster relief but then stayed on after other disaster relief teams had left to provide preventive and curative care. Data for the medical mission to Pakistan indicate that, right after the earthquake, Cuba sent a team of highly experienced disaster-relief specialists comprising 2,564 doctors (57% of the team), nurses, and medical technicians (Embassy of Cuba in Nueva Zealand 2010). Part of the team worked in refugee camps and Pakistani hospitals. Working in 30 field hospitals located across the earthquake-stricken zone, the team brought everything it would need to establish, equip, and run those hospitals. The cost to Cuba was not insignificant. Two of the hospitals alone cost US$500,000 each. In May 2006, Cuba augmented its aid with 54 emergency electrical generators.

Over the years, Cuba also has provided disaster relief aid to Armenia, Iran, Turkey, Russia, Ukraine, Belorussia, and most Latin American and Caribbean countries that have suffered either natural or man-made disasters. For example, over almost two decades, Cuba has treated free of charge almost 20,000 children - more than 16,000 Ukrainians, almost 3,000 Russians, and 671 Belorussians - mainly for post-Chernobyl radiation-related illnesses ('Cuba has treated' 2009). This type of medical diplomacy in the affected country's time of need has garnered considerable bilateral and multilateral symbolic capital for Havana, particularly when the aid is sent to countries considered more developed than Cuba.

Direct provision of medical care: selected examples

The Cuba-Venezuela-Bolivia connection: comprehensive health programs

It is indeed ironic that, in 1959, Fidel Castro unsuccessfully sought financial support and oil from Venezuelan president Rómulo Betancourt. It would take 40 years and many economic difficulties before another Venezuelan president, Hugo Chávez, would provide the preferential trade, credit, aid, and investment that the Cuban economy desperately needed. This partnership is part of the *Alternativa Bolivariana para los Pueblos de Nuestra América* (ALBA) to unite and integrate Latin America in a social justice-oriented trade and aid block under Venezuela's lead. Despite Fidel's three-decade-long obsession with making Cuba into a world medical power, ALBA also has created an opportunity to expand the reach of Cuba's medical diplomacy well beyond anything previously imaginable (Feinsilver 2006: 84, 2008: 111).

Cuba's current medical cooperation program with Venezuela is by far the largest it has ever attempted. These oil-for-doctors trade agreements allow for the preferential pricing of Cuba's exportation of professional services vis-à-vis a steady supply of Venezuelan oil, joint investments in strategically important sectors for both countries, and the provision of credit. In exchange, Cuba not only provides medical services to unserved and underserved communities in Venezuela - the initial agreement for massive medical services exports in 2005 was for 30,000 medical professionals, 600 comprehensive health clinics, 600 rehabilitation and physical therapy centers, 35 high-technology diagnostic centers, 100,000 ophthalmologic surgeries, and so on - but also provides similar medical services in Bolivia on a smaller scale at Venezuela's expense (Feinsilver 2006, 2008).

A later agreement included the expansion of the Venezuela-financed Cuban ophthalmologic surgery program – Operation Miracle – to perform 600,000 eyesight saving and restoration operations in Latin America and the Caribbean over a ten-year period. That number was surpassed already in late 2007, when the one-millionth patient was operated on. As of February 2010, 1.8 million patients had benefited from the program ('The open eyes' 2010). To achieve these numbers, Cuba established 61 small eye-surgery clinics in Venezuela, Bolivia, Ecuador, Guatemala, Haiti, Honduras, Panama, Nicaragua, Paraguay, Uruguay, Peru, St. Lucia, St. Vincent, Suriname, and Argentina; and Cuba extended the program to Africa, establishing clinics in Angola and Mali, to handle some of the demand from those and neighboring countries and to reduce the strain on facilities at home ('Nuevo impulso' 2009).

The second-largest medical cooperation program is with Bolivia, where in June 2006, 1,100 Cuban doctors were providing free health care, particularly in rural areas, in 188 municipalities. By July 2008, Cuban health personnel worked in 215 of Bolivia's 327 municipalities, including remote rural villages. It was reported that over the two-year period of medical diplomacy in Bolivia, Cuban doctors had saved 14,000 lives; had conducted more than 15 million medical exams; and had performed eye surgery on approximately 266,000 Bolivians and their neighbors from Argentina, Brazil, Paraguay, and Peru, the latter as part of Operation Miracle ('Report on Cuba' 2008). Ironically, through the Operation Miracle program in Bolivia, Cuba saved the eyesight of Mario Terán, the Bolivian man who killed Che Guevara.

Other Latin American and Caribbean examples

Cuban medical teams had worked in Guyana and Nicaragua in the 1970s, but by 2005, they were implementing the Comprehensive Health Program in Belize, Bolivia, Dominica, Guatemala, Haiti, Honduras, Nicaragua, and Paraguay. They also had established two comprehensive diagnostic centers, one on the island of Dominica and one on Antigua and Barbuda. Both Jamaica and Suriname's health systems are being bolstered by the presence of Cuban medical personnel, and the latter is implementing the Comprehensive Health Program ('Cuba coopera'; 'Cubans to help' 2008). Throughout the years, Cuba also has provided free medical care in its hospitals for individuals from all over Latin America and not just for the Latin American left.

Medical diplomacy beyond the Western Hemisphere

Cuba dispatched large civilian aid programs in Africa to complement its military support to Angola and the Horn of Africa in the 1970s and early 1980s. With the withdrawal of troops and the later geopolitical and economic changes of the late 1980s and the 1990s, Cuba's program remained but was scaled back (Feinsilver 1993: 156-195). Having suffered a postapartheid brain drain - white flight - South Africa began importing Cuban doctors in 1996. Already in 1998, 400 Cuban doctors practiced medicine in townships and rural areas, and in 2008, their number had increased slightly to 435. Cuban doctors began working in the Gambia in 1996, and since then and through 2009, 1,034 doctors, nurses, and medical technicians have served there (Jallow 2009). By 2004, there were about 1,200 Cuban doctors working in other African countries, such as Angola, Botswana, Cape Verde, Côte d'Ivoire, Equatorial Guinea, Gambia, Ghana, Guinea, Guinea-Bissau, Mozambique, Namibia, Seychelles, Zambia, Zimbabwe, and areas in the Sahara. By December 2005, Cuba was implementing its Comprehensive Health Program in Botswana, Burkina Faso, Burundi, Chad, Equatorial Guinea, Eritrea, Gabon, Gambia, Ghana, Guinea-Bissau, Guinea-Conakry, Mali, Namibia, Niger, Rwanda, Sierra Leone, Swaziland, and Zimbabwe.

On the African continent, South Africa is the financier of some Cuban medical missions in third countries. This South African-Cuban alliance has been much more limited in scope than the Venezuelan-Cuban deal. An agreement to extend Cuban medical aid into the rest of the African continent and a trilateral agreement to deploy more than one hundred Cuban doctors in Mali with US$1 million of South African financing were concluded in 2004. In January 2010, another South-South cooperation agreement was concluded for South African financing, also totaling $1 million, to support 31 Cuban medical specialists who already had been working in Rwanda for a year and had treated 461,000 patients (Karuhanga 2010).

Cuban medical teams also have worked and are working in such far-flung places as Timor-Leste (East Timor) in Southeast Asia and the Pacific island countries of Nauru, Vanuatu, Kiribati, Tuvalu, and the Solomon Islands, none of which might be considered in Cuba's strategic areas of interest. However, with one nation, one vote in the UN General Assembly, even these small islands are important where voting is concerned. The medical cooperation program in Timor-Leste began in December 2003 with the objective of creating a sustainable health care system by establishing the Cuban-model Comprehensive Health Program. In 2008, 177 medical professionals were providing a variety of services in Cuba's Comprehensive Health Program there ('Cuba coopera' 2008).

Although the actual numbers of Cuban doctors working in the Pacific islands is small, their impact is great. For example, when Cuba sent 11 doctors to the island of Nauru in September 2004, it provided 78% of all doctors in Nauru, an increase of 367% ('Cuban doctors help alleviate' 2004). Two Cuban doctors were working in the Solomon Islands in 2008, and the remaining seven arrived in early 2009 ('Cuban doctors' arrival' 2008). Three Cuban doctors currently work in Tuvalu, the first of whom arrived in October 2008. As of February 2009, they had attended 3,496 patients and saved 53 lives ('Cuban doctors inaugurated' 2009). Vanuatu and Cuba signed an agreement in 2008 for six Cuban doctors to work in provincial hospitals. Vanuatu's Director of Public Health Len Tarivonda indicated that his country would pay for return airfare and provide accommodations and a small local allowance, whereas the Cuban government paid the doctors' salaries. At that rate, he said: "Cuban doctors cost less than those from Australia and New Zealand" ('Vanuatu' 2008).

Medical Education

To contribute to the sustainability of other countries' health programs, Cuba has long provided full scholarships for foreign students to study medicine, nursing, dentistry, and medical technicians' courses in Cuba and has provided on-the-job training abroad. Likewise, Cuba has assisted other countries in establishing and staffing their own medical schools. However, it was not until 1999, the year after Hurricanes Mitch and Georges struck Central America and Haiti, that Cuba created the *Escuela Latinoamericana de Medicina* (ELAM) to provide personnel for the rehabilitation of those countries' health systems. Enrollment, however, was not limited to the affected countries. The six-year medical school program is provided free for low-income students who commit to practice medicine in underserved communities in their home countries on graduation. As part of the Cuba-Venezuela cooperation accords, Cuba agreed to train 40,000 doctors and 5,000 health care workers in Venezuela and provide full medical scholarships to Cuban medical schools for 10,000 Venezuelan medical and nursing students. In addition, Cuba offered Bolivia 5,000 more full scholarships to educate doctors and specialists as well as other health personnel at the ELAM in Havana. In 2006, there were some 500 young Bolivians studying at the school – about 22% of the total foreign scholarship student body – and another 2,000 had started the premed course there.

During the ELAM'S first graduation in August 2005, Hugo Chávez announced that Venezuela would establish a second Latin American Medical School so that, jointly with Cuba, the two countries would be able to provide free medical training to at least 100,000 physicians for developing countries over the next ten years. This led Cuba to implement the tutorial method of training medical personnel, whereby as of mid-2005, 12,000 Cuban doctors serving in the *Barrio Adentro* program in Venezuela became tutors for some 10,000 Venezuelan medical students. In March 2009, approximately 26,000 Venezuelan medical students were studying in the first four years of medical training as part of this new program to train comprehensive community doctors (*médicos integrales comunitarios*). This educational modality has been extended to six other countries in Asia, Africa, and Latin America, and another 14,000 students from 36 countries are studying under this modality in Cuba itself. Furthermore, in the 2008-2009 academic year, more than 24,000 foreign medical students were studying medicine at the ELAM (more than 7,900) and other Cuban institutions (more than 14,000) ('Cuba coopera').

In 2008, Cuba offered full medical school scholarships for 800 East Timorese students to begin work on the sustainability of their health system, of which 697 were studying in medical schools in Cuba and another 105 were studying under Cuban medical professors in East Timor (Anderson 2008). In 2009, in one medical faculty alone in the town of Sandino in western Cuba, there were a number of students from various South Pacific Islands and Timor-Leste studying medicine with full Cuban government scholarships: 199 from Timor-Leste, 50 from the Solomon Islands, 20 from Kiribati, ten from Tuvalu, seven from Nauru, and 17 from Vanuatu (Anderson 2009). Also in 2009, more than 300 nursing students from the English-speaking Caribbean, and two from China, participated in the Cuba-Caribbean Community (CARI-COM) training program for the provision of services to HIV/AIDS patients ('Cuba coopera'). During the 2009-2010 academic year, Cuba was training 51,648 medical students either in Cuba or in their own countries under the tutelage of Cuban professors. Of that number, 8,170 are enrolled in ELAM, 12,017 in the new program to train doctors (polyclinic based), 29,171 are being trained by Cuban medical brigades abroad, 1,118 are matriculated under other projects, and 1,172 are studying medical technician careers (Rojas Ochoa 2010).

The humanitarian benefits of this effort are enormous but so are the symbolic ones – prestige, influence, and goodwill – created. Moreover, the political benefits could be reaped for years to come as students trained by Cuba, with Venezuelan support, become health officials and opinion leaders in their own countries. Today, some of the 50,000 foreign scholarship students who trained in Cuban universities since 1961 (11,811 as doctors) are now in positions of authority and increasing responsibility (Prensa Latina 2008; Rojas Ochoa 2010).[3]

The costs and risks of medical diplomacy

The costs for beneficiary countries are relatively low. In most cases, the Cuban government pays doctors' salaries and the host country pays for airfare, room and board, and stipends of approximately $150-$375 per month depending on the country. This is far less than the costs of recruitment in the international marketplace, although it can still be a strain for cash-strapped economies. Perhaps more significant are the nonmonetary costs and risks involved. Cuban doctors serve the poor in areas in which no local doctor would work, make house calls a routine part of their medical practice, live in the neighborhood, and are available free of charge 24/7. This is changing the nature of doctor-patient relations and patients' expectations in the host countries. As a result, the presence of Cuban doctors has forced the reexamination of societal values and, in some cases, the structure and functioning of the health systems and the medical profession within the countries to which they were sent and where they continue to practice.

In countries such as Bolivia and Venezuela, although Cuban doctors generally are employed in areas where there are no local doctors, this different way of doing business has resulted in strikes and other protest actions by the local medical associations, as they are threatened by these changes and by what they perceive to be competition for their jobs. In the English-speaking Caribbean and in some Latin American countries, most particularly in Trinidad and Tobago, local medical associations have protested the different registration or accreditation standards applied to them and those applied or not applied to the Cuban doctors. Moreover, in Trinidad and Tobago, unlike other places where Cuban doctors serve, they pose a very real threat to local physicians' jobs because they were brought in by the government to fill vacancies left by both striking doctors and an overall insufficient number of local physicians to meet the country's health needs (Pickford-Gordon 2009).

The costs for Cuba, however, are more complicated partly because of the government's long-term investment in the education of medical personnel. Although Cuba pays the doctors' salaries, the pay scale is low by relative and absolute standards. In Cuba, doctors earn the equivalent in Cuban pesos of about US$25 per month. When they are abroad, that amount ascends to around US$185 per month. Since the Venezuelan agreement began, a significant amount of the costs for Cuba are, in fact, covered by Venezuela both for medical services and education for and in Venezuela and that provided to third countries. Previously, Cuba had fully funded these. However, money is fungible, and any aid Cuba receives could be channeled to this area.

A recent added cost has been the state's investment in the education and development of professionals who defect from medical diplomacy programs in third countries. Material conditions of life in Cuba are very difficult, and salaries are a fraction of those that can be earned abroad. These, as well as other factors, have enticed an estimated 900 to 2,000 medical professionals, not all doctors, to defect to the United States with a little stimulus from Uncle Sam.[4] In August 2006, the U.S. government announced the Cuban Medical Professional Parole Program, which grants Cuban doctors serving abroad fast-track asylum processing and almost-guaranteed entry into the United States. Although this program has encouraged more defections and even has provided a reason for some Cuban doctors to go abroad in the first place, some have found that they are held in limbo in Colombia or other points of arrival without the promised fast-track visa approval and with little or no money (Ceasar 2007). Others have made their way to the United States only to find that they cannot practice their profession because they must first pass the same four exams taken by U.S. medical graduates, but with the handicap of doing so in a foreign language in which they did not study medicine. And the focus of their medical education, primary care, and access to the latest technology differs from that of the United States. Cuban doctors also are older when they come to the United States and, thus, may have family responsibilities that could preclude their taking preparatory courses and studying for the exams instead of working in whatever jobs they might find (Ojito 2009).

A further risk for Cuba is increased dissatisfaction on the part of its own population as medical staff goes abroad, leaving some local health facilities and programs with insufficient staff despite the impressive ratio of doctors to population. As a result, a population accustomed to having a doctor on every block is finding that waiting times are now longer; medicines and supplies are scarcer; and where doctors are overworked, the quality of care declines.[5] Recognizing this problem in April 2008, Raúl Castro announced a reorganization of the Family Doctor Program at home to create greater efficiency by rationalizing the number and dispersion of family doctor offices but increasing the hours of operation for those outside of Havana until sufficient staff would become available. A year later, he announced further rationalization and cost containment with definite declines in both health care and education spending. Given the government's own proclamations that the health of the individual is a metaphor for the health of the body politic and that health indicators are a measure of government efficacy (Ministerio de Salud Pública 1983: 35 qtd in Feinsilver 1993: 1, 217), this situation could contribute to a delegitimization of the regime if insufficient attention is paid to the domestic health system.

Benefits of medical diplomacy

The value of Cuban medical diplomacy for the beneficiaries is clear. Over the past 50 years, Cuba's conduct of medical diplomacy has improved the health of the less privileged in developing countries while improving relations with their governments. Since 1961, Cuba has conducted medical diplomacy with 107 countries, deploying 134,849 medical professionals abroad, the large majority of whom were doctors ('Cuba coopera'; Rojas Ochoa). In April 2008, more than 30,000 Cuban medical personnel were collaborating in 74 countries around the globe (Prensa Latina 2008). Data as of October 2009[6] indicate that more than 37,000 Cuban medical professionals were deployed in 98 countries and four overseas territories ('Cuba coopera'). Overall, Cuban data show that, as of February 2009, Cuba's medical personnel abroad have saved more than 1.97 million lives, treated more than 130 million patients (of whom more than 39 million were seen on "house calls" at the patients' homes, schools, jobs, and so on), performed more than 2.97 million surgeries, and vaccinated with complete dosages more than 9.8 million people ('Cuba coopera'). Added to this are the previously mentioned 1.8 million sight-restoring and preserving eye surgeries conducted under Operation Miracle. Consequently, Cuban medical aid has affected the lives of millions of people in developing countries each year.

To make this effort more sustainable, over the years, more than 12,000 developing-country medical personnel have received free education and training in Cuba, and many more have benefited from education by Cuban specialists engaged in on-the-job training courses and/or medical schools in their own countries. In the largest enrollment ever, more than 50,000 developing-country scholarship students – and a small number of less-privileged Americans – were studying either in Cuban medical schools or under Cuban professors in their home countries during the 2009-2010 academic year. Furthermore, Cuba has not missed a single opportunity to offer and supply disaster-relief assistance irrespective of whether Cuba had good relations with that government. In fact, when Cuba established ELAM to help hurricane-ravaged Central American and Caribbean countries strengthen their health systems, none of the beneficiary governments was particularly friendly toward Cuba. In a more astonishing example, Cuba offered to send more than 1,000 doctors trained in disaster relief as well as medical supplies to the United States in the immediate aftermath of Hurricane Katrina. Although the Bush administration chose not to accept the offer, the symbolism of this offer of help by a small, developing country that has suffered 50 years of U.S. hostilities, including an economic embargo, is remarkable (Feinsilver 2006).

Since Cuba first sent a medical brigade to Chile in 1960, it has used medical diplomacy both to improve the health and win the hearts and minds of aid recipients and to improve relations with their governments.[6] Medical diplomacy has been a critical means of gaining symbolic capital - prestige, influence, and goodwill - which can translate into diplomatic support and material capital, such as trade or aid. It has been a way to project Cuba's image abroad as increasingly more developed and technologically sophisticated. More important, the practice of medical diplomacy also projects an image of Cuba as righteous, just, and morally superior because it is sending doctors rather than soldiers to far-flung places around the world. This latter comparison is important in Cuba's symbolic struggle as David versus the Goliath of the United States.

Cuba's success in this endeavor has been recognized by the WHO and other UN bodies, as well as by numerous governments, 107 of which have been direct beneficiaries of Cuba's medical largesse. It also has contributed to support for Cuba and rebuke of the United States in the UN General Assembly, where for the past 18 consecutive years members voted overwhelmingly in favor of lifting the U.S. embargo of Cuba. In fact, only Israel and Palau have supported the U.S. position, and the Marshall Islands and Micronesia abstained (see for example Nichols 2009). With equal voting rights for all members of the UN General Assembly, Cuba's medical diplomacy with such a large number of member states is a rational endeavor, however humanitarian the impetus may be.

Furthermore, the success of Cuba's medical diplomacy was made evident once again at the Summit of the Americas in Trinidad in April 2009. The Latin American heads of state frequently mentioned it in their discussions with President Obama. Cuba's medical diplomacy underpins their support for lifting the U.S. trade embargo on Cuba and normalizing relations, including the reinstatement of Cuba into the Organization of American States (OAS), agreed to in the June 2009 OAS meeting in Honduras, although Raúl Castro indicated that Cuba was disinterested. In turn, Obama mentioned this fact in his own remarks, even indicating that the United States could learn from Cuba. He has been widely quoted as later saying, "We have to use our diplomatic and our development aid in more intelligent ways so that people can see the very practical, concrete improvements in the lives of ordinary persons as a consequence of U.S. foreign policy" (Carlsen 2009; Blanchfield 2009).

Economic benefits have been very significant since the rise of Chávez in Venezuela. Trade with and aid from Venezuela in a large-scale oil-for-doctors exchange have bolstered Cuba's ability to conduct medical diplomacy and, importantly, have helped keep its economy afloat. Earnings from medical services, including the export of doctors, equaled 28% of total export receipts and net capital payments in 2006. This amounted to US$2.312 billion, a figure greater than that for both nickel and cobalt exports and tourism (Embassy of India (Havana) 2007).[7] In fact, the export of medical services is thought to be the brightest spot on Cuba's economic horizon (Mesa-Lago and Ritter

2008). Data for 2008 demonstrates that Cuba earned about US$5.6 billion for the provision of all services to Venezuela, most of which were medical, although the figure includes teachers and other professionals. The total value of the Venezuelan trade, aid, investments, and subsidies to Cuba for 2008 was US$9.4 billion (Mesa-Lago forthcoming).

Medical diplomacy also paves the way for Cuba's export of a range of medical products. In this context, for example, Cuban exports of medicines to ALBA countries increased by 22% from 2008 to 2009 ('Más medicamentos' 2010). It is quite likely that other countries receiving Cuban doctors will also purchase Cuban vaccines, medicines, medical supplies, and equipment. Cuba's biotech industry holds 1,200 international patents and earned US$350 million in product sales in 2008 (Grogg 2009; Shafrin 2009). Potential for growth in the export of vaccines is good, particularly in joint ventures with other countries, such as Cuba has already with Brazil and China, and with big pharmaceutical companies, like GlaxoSmithKline (Yee 2003).

Symbolic capital garnered from both the success of the domestic health system and medical diplomacy made possible Cuba's establishment of a medical tourism industry. Although begun as a small program in 1980, medical tourism became important by 1990, with the collapse of the Soviet Union and after Cuba had vastly increased its production of doctors for medical diplomacy programs. The number of patients participating in medical tourism in Cuba for the first eight years of the program was equal to 55% of medical tourists in 1990 alone. Revenue from health tourism in 1990 was US$ two million (Feinsilver 1993: 190). This program received renewed impetus during the mid-1990s as the government sought to increase its foreign exchange earnings through a variety of methods, including limited foreign investment. By 1997, revenue had increased to US$20 million, 98.5% of which was plowed back into the domestic health system (Brotherton 2008).

On the domestic front, medical diplomacy has provided an escape valve for disgruntled medical professionals who earn much less at home than less skilled workers in the tourism sector. Their earning potential is much greater abroad, both in the confines of the medical diplomacy program and even more so beyond it. This constant lure of defection has led the Ministry of Public Health to establish a coefficient for possible defections - two to three percent of the total number of international medical collaborators - as part of precise human-resources-planning exercises (Anonymous 2009). Moreover, medical diplomacy has given Cuban doctors and other medical personnel an opportunity to bring home from their deployment station consumer goods unavailable in Cuba. In this way, it is has helped defuse the tension between the moral incentives of socialist ideology and the material needs of Cuba's decidedly hardworking and no-less-dedicated medical personnel.[8]

Conclusion

Medical diplomacy has been a cornerstone of Cuban foreign policy since the outset of the revolution 50 years ago. It has been an integral part of almost all bilateral relations agreements that Cuba has made with other developing countries. As a result, Cuba has positively affected the lives of millions of people per year through the provision of medical aid, as well as tens of thousands of foreign students who receive full scholarships to study medicine either in Cuba or in their own countries under Cuban professors. At the same time, Cuba's conduct of medical diplomacy with countries whose governments had not been sympathetic to the revolution, such as Pakistan, Guatemala, Honduras, and El Salvador, to name only a few, has led to improved relations with those countries.

Medical diplomacy has helped Cuba garner symbolic capital (goodwill, influence and prestige) well beyond what would have been possible for a small, developing country, and it has contributed to making Cuba a player on the world stage. In recent years, medical diplomacy has been instrumental in providing considerable material capital (aid, credit, and trade), as the oil-for-doctors deals with Venezuela demonstrates. This has helped keep the revolution afloat in trying economic times.

What began as the implementation of one of the core values of the revolution, namely health as a basic human right for all peoples, has continued as both an idealistic and a pragmatic pursuit. As early as 1978, Fidel Castro argued that there were insufficient doctors to meet demand in the developing world, despite the requesting countries' ability to pay hard currency for their services (Bohemia, cited in Feinsilver 1993: 264 n201). Because Cuba charged less than other countries, with the exception at that time of China, it appeared that it would win contracts on a competitive basis. In fact, during the following decade (1980s), Cuba's medical contracts and grant aid increased. In most cases, aid led to trade, if not to considerable income. With the debt crises and the International Monetary Fund's structural adjustment programs of the 1980s, grant aid predominated. In 1990, Cuban medical aid began to dwindle as neither the host countries nor Cuba could afford the costs, the former because of structural adjustment-mandated cuts in social expenditures and the latter because of the collapse of its preferential trade relationships following the demise of the Soviet Union. As Cuba's ability to provide bilateral medical aid diminished, its provision of medical aid through multilateral sources (contracts) increased (Feinsilver 1993: 193-194). Cuba's medical diplomacy continued, albeit on a smaller scale during the 1990s, until the rise of Hugo Chávez in Venezuela.

With medical services leading economic growth in the twenty-first century, it seems unlikely that even the more pragmatic Raúl Castro will change direction now. In contrast, dependency on one major benefactor and/or trade partner can be perilous, as the Cubans have seen more than once. If Chávez either loses power or drastically reduces foreign aid in an effort to cope with Venezuela's own deteriorating economic conditions and political opposition, Cuba could experience an economic collapse similar to that of the Special Period in the early 1990s. In fact, the global financial and economic crisis has compounded existing problems. In an effort to avert that type of collapse, Raúl Castro has been trying to further diversify Cuba's commercial partners (Mesa-Lago forthcoming). In July 2009, Cuba received a new US$150 million credit line from Russia to facilitate technical assistance from that country, and companies from both countries signed various agreements, including four related to oil exploration ('Cuba y Federación Rusa' 2009). Furthermore, Raúl Castro made clear in his August 1, 2009, speech before the National Assembly of People's Power, that Cuba could not spend more than it made (Castro Ruz 2009). He asserted that it was imperative to prioritize activities and expenditures to achieve results, overall greater efficiency, and to rationalize state subsidies to the population.

Despite a little help from their Venezuelan friend, the Cuban government has had to embark on austerity measures that hark back to the worst of times right after the collapse of the former Soviet Union (Mesa-Lago 2009). With two budget cuts already this year, restrictions on electricity distribution, and a 20% decrease in imports (Foreign Staff 2009; 'Cuba sounds energy alarm' 2009), it is likely that the Cuban government will attempt to increase its medical exports to countries that can afford to pay for them. In fact, in August 2009, Raúl Castro indicated that Cuba would need to increase the production of services that earn hard currency.[9] Pragmatism clearly dictates this course of action even if it also is imbued with strong revolutionary idealism about humanitarian assistance.

Economic and political benefits of medical diplomacy aside, Fidel, both when he was president and today as an elder statesman and blogger, most sincerely cares about health for all, not just for Cubans. His long-term constant involvement in the evolution both of the domestic health system and of medical diplomacy has been clear through both his public pronouncements and actions, and the observations and commentary of his subordinates and external observers (Feinsilver 1993; Bourne 1986: 284). Today, this concern for health is part of the social agenda of ALBA, through which, for example, additional Cuban medical aid to Haiti post-2010 earthquake is being conducted.

Unable to offer financial support, Cuba provides what it excels at and what is easily available, its medical human resources. International recognition for Cuba's health expertise has made medical diplomacy an important foreign policy tool that other, richer countries would do well to emulate. After all, what country could refuse humanitarian aid that for all intents and purposes appears to be truly altruistic?

References

Andaya, E. (2009) 'The gift of health: socialist medical practice and shifting material and moral economies in post-Soviet Cuba', *Medical Anthropology Quarterly*, 23.4: 357-74.

Anderson, T. (2008) *The Doctors of Tomorrow: The Timor Leste-Cuba Health Cooperation.* (Documentary) University of Sydney.

— (2009) *The Pacific School of Medicine.* (Documentary) University of Sydney. Online. Available HTTP: <http://www.youtube.com/watch?v=AhMAncnEDQQ> and <http://www.youtube.com/watch?v=j-AoKYCDmlo&feature=related>.

Anonymous (2009, April 30). *Interview with anonymous source #25.*

Blanchfield, M. (2009, April 19) 'Harper hails unplugging of ideological megaphones at hemispheric summit', *Canwest News Service.* Online. Available HTTP: <http://www.canada.com/business/Harper+hails+unplugging+ideological+megaphones+hemispheric+summit/1512580/story.html>.

Bohemia (1978, September 15). Cited in Feinsilver, *Healing the Masses.*

Bourne, P.G. (1986) *Fidel: A Biography of Fidel Castro*, New York: Dodd, Mead.

Brotherton, P.S. (2005) 'Macroeconomic Change and the Biopolitics of Health in Cuba's Special Period', *Journal of Latin American Anthropology*, 10.2: 339-69.

— (2008) '"We have to think like capitalists but continue being socialists":
medicalized subjectivities, emergent capital, and socialist entrepreneurs
in post-Soviet Cuba', *American Ethnologist*, 35.2: 259-74.

Carlsen, L. (2009, April 22) 'Words and deeds in Trinidad', *Foreign Policy in
Focus*. Online. Available HTTP: <http://www.fpif.org/articles/words_
and_deeds_in_trinidad>.

Castro Ruz, R. (2009, August 1) 'Year 52 of the 50th Anniversary of the
Revolutionary Triumph', speech presented at the 3rd Regular Session of
the Seventh Legislature of the National Assembly of People's Power,
Havana, Cuba. Online. Available HTTP: <http://en.cubadebate.cu/
opinions/2009/08/01/speech-3rd-regular-session-seventh-legislature-
national-assembly-peoples-power/>.

Ceasar, M. (2007, August 1) 'Cuban doctors abroad helped to defect by new
U.S. visa policy', *World Politics Review*.

'Cuba considera ayuda a Haiti prioritaria', (2010, January 13) *Granma*. Online.
Available HTTP: <http://www.granma.cubaweb.cu/2010/01/13/
nacional/artic28.html>.

'Cuba coopera' (2008). Online. Available HTTP: <http://www.cubacoop.
com/cubacoop/Cooperacion_ResultadosG>, (accessed 12 October
2009).

'Cuba has treated over 20,000 children from Chernobyl disaster', (2009, April
2) *Havana Journal*. Online. Available HTTP: <http://havanajournal.com/
forums/viewthread/1145>.

'Cuba sounds energy alarm, plans blackouts', (2009, May 26). *Associated Press*.
Online. Available HTTP: <http://www.azcentral.com/news/
articles/2009/05/26/20090526cuba-blackouts0526-ON.html>.

'Cuba y Federación Rusa estrechan colaboración económica', (2009, July 29)
Granma Internacional. Online. Available HTTP: <http://www.granma.cu/
espanol/2009/julio/mier29/colaboracion.html>.

'Cuban doctors' arrival a blessing, says Solomon's health dept.' (2008, June
10) *Radio New Zealand International*. Online. Available HTTP: <http://
www.rnzi.com/pages/news.php?op=read&id=40275>.

'Cuban doctors help alleviate Nauru health problems', (2004, September 7)
ABC News Online. Online. Available HTTP: <http://www.abc.net.
au/news/2004-09-07/cuban-doctors-help-alleviate-nauru-health-
problems/2039848>.

'Cuban doctors in Haiti boost integral assistance' (2010, February 2) *Prensa
Latina*. Online. Available HTTP: <http://www.solvision.co.cu/english/

index.php?option=com_content&view=article&id=862:cuba-doctors-in-haiti-boost-integral-assistance-&catid=7:health&Itemid=128>.

'Cuba doctors inaugurated new health services in Tuvalu, a small pacific island' (2009, June 14) *Cuba Headlines*. Online. Available HTTP: <http://www.cubaheadlines.com/2009/06/14/17664/cuban_doctors_have_inaugurated_a_series_new_health_services_tuvalu_a_small_island_nation_pacific.html>.

'Cubans to help boost local health sector', (2008, May 10) *Jamaica Observer*. Online. Available HTTP: <http://www.jamaicaobserver.com/news/135451_Cubans-to-help-boost-local-health-sector>.

Embassy of Cuba in Nueva Zealanda. (2010, January 22) 'Cuba stands by the Haitian people', *Pacific Press Release*. Online. Available HTTP: <http://pacific.scoop.co.nz/2010/01/cuba-stands-by-the-haitian-people/>.

Embassy of India (Havana), (2007, April 13) 'Annual commercial and economic report - 2006'.

Feinsilver, J. (1989a) 'Cuba as a "world medical power": the politics of symbolism', *Latin American Research Review*, 24.2: 1-34.

— (1989b) 'Cuban Medical Diplomacy', paper presented at Latin American Studies Association meeting, Halifax Nova Scotia, November 1989.

— (1993) *Healing the Masses: Cuban Health Politics at Home and Abroad*, Berkeley, CA: University of California Press.

— (2006) 'La diplomacia médica cubana: cuando la izquierda lo ha hecho bien', *Foreign Affairs en Español*, 6.4: 81-94. Online. Available HTTP <http://www.coha.org/2006/10/30/cuban-medical-diplomacy-when-the-left-has-got-it-right/>.

— (2008) 'Médicos por petroleo: la diplomacia médica cubana reciba una pequeña ayuda de sus amigos', *Nueva Sociedad* (Buenos Aires), 216: 107-22.

— (2009a) 'Cuba's Health Politics at Home and Abroad', in L. Panitch and C. Leys (eds.) *Morbid Symptoms: Health under Capitalism, Socialist Register 2010*, London: Merlin Press.

— (2009b) 'Cuban Medical Diplomacy', in M. Font (compiler) *A Changing Cuba in a Changing World*, New York: Bildner Center, City University of New York.

Foreign Staff (2009, August 9) 'Cuba runs out of lavatory paper', *The Telegraph*. Online. Available HTTP: <http://www.telegraph.co.uk/news/newstopics/howaboutthat/6001005/Cuba-runs-out-of-lavatory-paper.html>.

Grogg, P. (2009, December 1) 'Welfare of Cuban people is bottomline of Cuba's pharmaceutical industry', *New Jersey Newsroom*. Online. Available HTTP: <http://www.newjerseynewsroom.com/international/welfare-of-the-people-is-bottomline-for-cubas-pharmaceutical-industry/page-2>.

Jallow, A. (2009, December 7) 'Gambia: 1034 Cuban medical personnel work in the country', *Banjul Daily Observer*. Online. Available HTTP: <http://allafrica.com/stories/200912080823.html>.

Karuhanga, J. (2010, January 22) 'MoH gets $1m to facilitate Cuban medical volunteers', *Rwanda New Times*. Online. Available HTTP: <http://www.newtimes.co.rw/news/index.php?i=14148&a=25078>.

'Más medicamentos cubanos para el ALBA', (2010, January 6) *Radio Sucre*. Online. Available HTTP: <http://www.radiosucre.cu/Salud.php?id=4786>.

Mesa-Lago, C. (2009, July 12) 'La paradoja económica cubaba', *El Pais*. Online. Available HTTP: <http://www.elpais.com/articulo/semana/paradoja/economica/cubana/elpepueconeg/20090712elpneglse_10/Tes>.

— (forthcoming) 'The Cuban economy in 2008-2009: internal and external challenges, state of the reforms and perspectives', in P. Spadoni (ed.). Online. Available HTTP: <http://stonecenter.tulane.edu/uploads/Mesa-Lago-1305235386.pdf>.

Mesa-Lago, C. and Ritter, A. 'Remarks', presented at City University of New York Bildner Center Conference *A Changing Cuba in a Changing World*, New York, March 2008.

Ministerio de Salud Pública, Republica de Cuba (1983) *Salud para todos: 25 años de experiencia cubana*, Havana: Ministerio de Salud Pública.

Nichols, M. (2009, October 28) 'U.N. votes against U.S. embargo on Cuba for 18th year', *Reuters*. Online. Available HTTP: <http://www.reuters.com/article/idUSTRE59R4LQ20091028/>.

'Nuevo impulso a Operación Milagro en Argentina' (2009, December 3) *Radio Sucro*.

Oficina Nacional de Estadísticas, Republica de Cuba. (2009) *Anuario estadístico de Cuba 2008: edición 2009*. Online. Available HTTP: <http://www.one.cu/aec2008/esp/20080618_tabla_cuadro.htm>.

Ojito, M. (2009, August 4) 'Doctors in Cuba start over in the US', *New York Times*. Online. Available HTTP: <http://www.nytimes.

com/2009/08/04/health/04cuba.html?pagewanted=1>.

Pickford-Gordon, L. (2009, August 9) 'Cuban doctors and TT's medical development', *Trinidad and Tobago's Newsday*. Online. Available HTTP: <http://www.newsday.co.tt>.

Prensa Latina (2008, April 11).

'Reflexiones del compañero Fidel: La cumbre secreta' (2009, April 21) *Diario Granma*. Online. Available HTTP: <http://www.granma. cubaweb.cu/secciones/ref-fidel/art125.html>.

'Report on Cuban healthcare professionals in Bolivia' (2008, July 16) *Periódico*. Online. Available HTTP: <http://www.periodico26.cu/ english/health/jun_sep2008/doctors-bolivia071608.html>.

Rigoli, F. (2009, April 30) *Interview*. Pan-American Health Organization.

Rojas Ochoa, F. (2010, February 3) *Personal Communication*.

Shafrin, J. (2009, March 26) 'Cuban exports: sugar, cigars...and cancer drugs?', *Healthcare Economist*. Online. Available HTTP: <http:// healthcare-economist.com/2009/03/26/cuban-exports-sugar-cigarsand-cancer-drugs/>.

'Solomon Islands welcome Cuba doctors', (2009, March 25) *Radio New Zealand International*. Online. Available HTTP: <http://www.rnzi.com/ pages/news.php?op=read&id=40275>.

'The open eyes of Latin America' (2010, February 3) *Granma*. Online. Available HTTP: <http://translate.google.com/translate?langpair=es |en&u=http%3A%2F%2Fwww.radioangulo.cu%2Fenglish%2Findex. php%3Foption%3Dcom_content%26task%3Dview%26id%3D5909%2 6Itemid%3D28>.

'Vanuatu to get six doctors from Cuba', (2008, August 10) *Radio New Zealand International*. Online. Available HTTP: <http://www.rnzi.com/pages/ news.php?op=read&id=41373>.

Yee, C.M. (2003, April 17) 'Cutting-edge biotech in old-world Cuba', *Christian Science Monitor*. Online. Available HTTP: <http://www. latinamericanstudies.org/cuba/biotech.htm>.

Endnotes

1   This chapter was originally published as Feinsilver, J.M. (2010) 'Fifty Years of Cuba's Medical Diplomacy', *Cuban Studies*, 41: 85-104, and is reprinted with permission.

2   Although the statistical data in the book are dated, the overall analysis of the whole book has, as Dr. Peter Bourne – executive producer of *Salud! The Film* – indicated in a personal communication of November 13, 2006, "withstood the test of time."

3   Data on medical graduates from 1963 to 2008 from personal communication from Dr. Francisco Rojas Ochoa, Distinguished Professor and Editor, *Revista Cubana de Salud Pública*, (3 February 2010).

4   Anti-Castro Cuban-American Representative Lincoln Diaz-Balart's (R-FL) chief of staff, Ana Carbonell, told Mirta Ojito of *The New York Times* that about two thousand Cuban medical professionals, not all doctors, had settled in the United States since the 2006 program began. However, I have not been able to corroborate these numbers despite attempts through various government agencies and officials. See Ojito 2009. The Cuban Ministry of Public Health planners calculate a defection rate of between two percent and three percent when planning their human resource needs (Rigoli 2009).

5   For some excellent discussions of the material problems facing Cuban doctors and their patients in post-Soviet Cuba based on anthropological field work, see Brotherton 2005; Andaya 2009.

6   In a recent article, Fidel Castro refuted the idea that Cuba has used medical diplomacy to gain influence. Nonetheless, the evidence suggests that it has done so, even though this might not have originally been the primary reason for doing so. It is an intelligent use of Cuba's comparative advantage, its medical human resources. See 'Reflexiones del compañero Fidel' 2009.

7   Embassy of India (Havana), "Annual Commercial and Economic Report - 2006," No.Hav/Comm/2007, April 13, 2007.

8   On the "economy of favors" or "the moral economy of ideal socialist medical practice… based on reciprocal social exchange," see Andaya 2009: 357.

9   Raúl Castro stated that Cuba will increase production of services that generate foreign exchange (see Castro Ruz 2009).

5

Fueling la Revolución: Itinerant Physicians, Transactional Humanitarianism, and Shifting Moral Economies in Post-Soviet Cuba

P. Sean Brotherton

> The cooperative efforts of Venezuela and Cuba, highlighted by their medical programs, are serving as the models for humanitarian relationships and fair economic exchange between the nations of the south.
> Steve Brouwer 2011: 230

> [T]he word humanitarianism tends to elide critical analysis. Because it is a valued good that many are trying to appropriate for themselves by qualifying their own activities as "humanitarian," even when they are warlike, and because it operates by internalizing debate on the meaning and effects of its actions, it resists the inquiry of social sciences.
> Didier Fassin 2010: 36

The Lived Experience of Socialism

As international health analysts decry widening global health inequities, deteriorating health conditions among populations in the global South have served as a moral barometer to measure the failure of existing humanitarian aid programs. Addressing global health problems has thus emerged as an ideological battleground for competing political imaginaries on how to redefine health within a social justice paradigm. In this fraught terrain, Cuban health professionals have assumed a prominent role given their medical education and training, which have been touted by Cuban officials and international policy analysts alike as the realization of Ernesto "Che" Guevara's vision of physicians as vehicles of social change (see, for example, Brower 2011). Contemporary iterations of Guevara's (1968) rallying call for physicians to embody revolutionary praxis, as espoused in his canonical *On Revolutionary Medicine*, can be found in the Cuban Ministry of Public Health (MINSAP) training manual, which declares that physicians should "abide by moral and ethical principles of deep human, ideological and patriotic content; such as, to dedicate their efforts and knowledge to the improvement of the health of humankind, to work constantly where society requires them and to always be available to offer necessary medical attention where it is needed" (MINSAP 2001: 10). As the physicians training manual further states, "The practice of the health professional has in itself a profound ethical underpinning that is geared toward promoting the highest values in human beings in order to pay homage every day, and in every sphere, to the work of la Revolución, of which we are children" (MINSAP 2001: 10).

The trope of the valiant family physician as revolutionary hero, at home or on missions abroad, has become a dominant leitmotif in numerous books, Cuban newspapers like Communist Party daily, *Granma*, and journals like *Bohemia*.[1] In these representations, physicians laboring out of heroic sacrifice are a testament to the importance of moral, as opposed to material, rewards for "curing the ills of society," as Guevara (1986) famously said. Over the last two decades, Cuba's rapid expansion of an ejército de batas blancas (army in white lab coats) traveling the world – often "going where no doctor has gone before" – has garnered accolades for transforming the landscape and, indeed, bodies in the political economy of health care delivery in Latin America and the Caribbean, as well as several African countries. Revolutionary physicians, then, have become the staple labor force in medical missions abroad.

Since the early 2000s, medical internationalism programs have developed and incrementally expanded through the provision of free humanitarian medical treatment to citizens of various countries throughout the world.[2] Many of these medical missions, and the sheer empirical volume of what has been accomplished to date, have been prominently featured in Cuban media outlets inside the country and highlighted in public relations campaigns conducted by Cuban embassies abroad. The analysis presented in the following pages is not intended as an affront to the laudable work of Cuban physicians. Instead, my aim is to pursue another line of inquiry. I follow the lead of anthropologist Didier Fassin (2010), among others, who argues that humanitarian efforts should not be immune from critical scrutiny because of claims that they "do good." In this chapter, I examine how the post-Cold War era witnessed the reconfiguration of global relations. This has meant that Cuba, as a result of its disadvantaged position, must increasingly forge strategic relationships that are contingent on capitalizing on their medical know-how and personnel; in effect, the state is making commodities (rather than gifts) out of the very things that have served as the symbols of its success. I wish to explore the domestic effects, both material and discursive, of Cuba's physicians being sent or, more specifically, "sold" to disparate locations throughout the world. Cuban officials label this program variously as "medical internationalism" or "medical cooperation." I argue that this is a form of transactional humanitarianism, that is, an assemblage of traveling actors, experts, practices, and specialized knowledge that are collectively marketed under the umbrella term "humanitarian," yet are ostensibly embedded in market relations and shifting moral values of exchange.

The language of economic incentives is a thorny issue in the now ubiquitous and growing field of humanitarianism. A robust body of scholarship in the social sciences has already highlighted the nexus among various forms of humanitarian aid, conditional trade agreements, militarism, and highly volatile flows of capital. For the most part, this literature has parsed out the ethical, economic, and ethnographic inner-workings of humanitarian aid programs and how they, at times, reproduce a liberal democratic order.[3] These programs, these scholars note, are traditionally funded by wealthy developed countries aimed at countries in the global South, as well as countries experiencing civil strife or natural disasters. Comprising a coalescence of both a "politics of life" and "interventionist regimes of governance," such humanitarian aid programs participate directly or indirectly in adjudicating, assessing, and evaluating who and what conditions produce victims, social suffering, and political instability, among other salient concerns. In doing so, these programs, ultimately, identify and delimit areas or issues that warrant intervention and set the very terms and conditions in which such assistance is offered.

In this context, how can we conceptualize the fact that a small, resource-poor nation such as Cuba has become a leading figure in delivering "humanitarian biomedicine"[4] to the world's poor? Does Cuba's project of "exporting doctors" – more recently, on a massive scale – constitute what anthropologist James Ferguson (2006: 26) describes as an "inconvenient case," that is, a case that stands in stark contrast to theories of global connection and ruptures the romantic portrayals of flows of capital, goods, commodities, and people. On the surface, the socialist humanitarian project eschews traditional binaries of capital flow from the North to the South. But, the question remains: is there another logic at work here? Cuba's medical internationalism programs require us to seriously ponder the implications of what it means to export la Revolución in the name of humanitarianism. The swelling number of "revolutionary physicians," has come to represent the "Cubanization" of public health reform in many countries in the Americas (see, for example, Bernstein's chapter in this volume), and this has not occurred without controversy or culture clashes. These issues merit entirely separate papers. What I seek to highlight in this chapter is how the very principles of la Revolución are themselves in flux, at times contradictory, often leading Cubans in Cuba to construct new meanings and interpretations of what it means to participate (or not) in the ongoing socialist project. This is an example of how changes in the everyday "lived experience of socialism" have led citizens to start critically questioning how the material benefits of strategic aid programs fail to redress domestic shortages, and even lead to worsening material circumstances for ordinary people. To put it more broadly, this issue can help us determine what la Revolución writ large means for Cubans in Cuba in an era rife with contradictions, scarcity, and the everyday struggle for basic necessities.

Internacionalistas: Itinerant Physicians as Agents of Social Change?

> And our task now is to orient the creative abilities of all medical professionals toward the tasks of social medicine.
>
> Ernesto Guevara, On Revolutionary Medicine

Hoping to avoid the oppressive humidity of the late summer months, I arrived in Havana in May 2004 to carry out several weeks of research. As I systematically made my way through several communities where I previously conducted ethnographic fieldwork (from 1999-2002), exchanging greetings with old friends, past interlocutors, and passing acquaintances, something was immediately obvious. Most of the primary health clinics, known as consultorios, I had previously worked in were now vacant. When I asked about the changes, people were quick to point out, with a sense of pride, that their respective physicians had been selected to participate as an internacionalista, a physician working in medical brigades outside of the country. Despite the everyday lived experience of sacrifice, struggle, and material scarcity, most of the residents of the affected consultorios discussed the country's ongoing humanitarian aid programs in admiring terms. Although an economically disadvantaged player in the global sphere, Cuba had desirable resources in the form of human capital and medical expertise, many people noted. The government's gifts of medical technology, aid, and health personnel to "deserving recipients," as many described, were a source of immense individual and national pride.

A former Cuban diplomat stationed in London who had worked in several Latin American countries cited an excellent example of the different way Cuba's medical missions are framed within competing nationalist discourses. The diplomat claimed that despite the involvement of certain governments – Nicaragua and Guatemala, for example – in siding with the United States on various foreign policies against Cuba, the Cuban government was committed to providing free medical assistance to those countries. As the diplomat concluded, Cuba wanted to demonstrate that socialism was about promoting solidarity, not about forcing people to embrace a political agenda against their will. Cuba used actions, not words, to demonstrate the power of socialism, he asserted. Political will was the reason that Cuba accomplished its goals. Furthermore, individuals in various countries who benefited from free medical aid come to understand the extent of Cuba's generosity on their own terms.

This humanitarian ethic has been part of the country's international solidarity campaigns since the early 1960s. For example, in 1960 Cuba sent its first medical team to Chile following an earthquake. Shortly thereafter, in 1963, Cuba launched its official program of "medical missions" by sending an international medical brigade to Algeria as part of its goal of supporting anticolonial struggles. In more recent memory, the medical brigades sent throughout Central America and the Caribbean after hurricanes George and Mitch in 1998, and the treatment of victims of the Chernobyl disaster in 1986 on the island, reflected the country's ethos of humanitarianism and commitment to internationalism. This culture of caring was unquestioned in the various conversations that ensued in my interviews with officials, physicians, and average citizens. The message was clear: Cuba played a prominent role on the world stage as a medical power. On one hand, Cuba was selfless in providing much-needed medical aid in times of need. On the other, foreign countries often sought out Cuban medical advice, training, and personnel in the express hope of replicating the Cuban model of primary health care delivery. In doing so, these countries also hoped to produce comparable health outcomes.

As was evident in many of the interviews I conducted in the late 1990s and early 2000s, Cubans had, in many ways, participated in and helped propagate, relish, and circulate the state's narrative of "international solidarity" as a moral imperative of the country's socialist vision. Years later, however, during fieldwork in 2007 and 2010, I noted a palpable shift in the popular discourse among everyday citizens on the rationale behind the increasing departure of

Figure 5.1: "Mission Inside the Neighborhood."
Source: http://www.barrioadentro.gov.ve/ (accessed 18 February 2012)

physicians, nurses, pharmacists, and even teachers, to go on medical missions. The discourse of humanitarianism was no longer at the forefront of many of these discussions. Rather, the focus was now, without fail, on *los Venezolanos (the Venezuelans)*. In March 2003, Barrio Adentro (Inside the Neighborhood) emerged out of an agreement between the governments of Venezuelan president Hugo Chávez and Castro (see Figure 5.1). Under this program, over 20,000 Cuban physicians and auxiliary health professionals would be stationed in primarily poor neighborhoods in Venezuela, providing medical care in exchange for highly subsidized petroleum – popularly dubbed the oil-for-aid deal. At the same time, the Cuban government also created an eye surgery program known as Operación Milagro, or Operation Miracle, which sought to treat over 200,000 patients in 21 countries with the aim of restoring sight lost to cataracts, glaucoma, diabetes, and other diseases (see Figure 5.2). For the most part, the countries participating in this program had agreed to pay for these services with subsidized trade agreements with Cuba.

For example, in 2007 one of my neighbors commented that Hugo Chávez had, in effect, become the country's "Petro Papi"; Venezuela's role was similar to the one the former USSR had played in providing highly specialized and dependent forms of trade and aid. Plan Barrio Adentro (later renamed Misión Barrio Adentro, or MBA) was established in Venezuela in 2003 as part of larger trade agreements; the first was signed in 2000 and the second in 2005.[5] These agreements, steeped in the language of social justice, were framed as a decisive move away from the hegemonic influence of neolib-

Figure 5.2: Mission Miracle: A Solitary Vision of the World
Source: http://www.barrioadentro.gov.ve/ (accessed 18 February 2012)

eralism. Eclipsed from this selective framing, however, was how the MBA program equally emerged out of "techniques of calculative choice institutionalized in mechanisms and procedures that mark[ed] out special spaces of labor markets, investment opportunities, and relative administrative freedom" (Ong 2006: 19). For instance, pivotal to Cuba's recent transactional humanitarian efforts in Latin America is the country's inclusion in strategic partnership agreements such as the Bolivarian Alliance for the Americas, known as ALBA. This partnership is grounded on the fundamental belief that true international exchange must be directed through states and public action rather than through individuals and private markets.

As Feinsilver's chapter in this volume adroitly examines, the economic benefits of such agreements for Cuba were great: preferential pricing of Cuba's exportation of professional services in return for a steady supply of Venezuelan oil, joint investments in strategically important sectors for both countries, and the provision of credit.[6] I assert that a distinctive variant of humanitarian aid is being deployed here, and it is envisioned as a strategic, if not essential, project of generating revenue and cultivating new markets of exchange. While itinerant physicians on medical missions throughout the Soviet period (1959-1989) could be couched in the language of symbols of the revolution and be morally justified, what are some of the domestic ramifications of Cuba's recent export of health personnel, or human capital en masse, to foreign locales? *As more Cuban physicians participate in strategic aid programs such as Barrio Adentro or Operación Milagro, many Cubans citizens are starting to ask why Venezuelans are more deserving of the gifts of Cuba's medical aid programs than Cubans themselves. Why is it so hard to find a family physician in a country with an apparent surplus?* To answer these questions, we must briefly touch on the historic role of Cuba's "health revolution" at home and, importantly, critically examine how, in recent years, the institutions of the state have increasingly modified their policies, objectives, and age-old ideological positions. This has influenced how medicine is practiced, experienced, and imagined in the post-Soviet era.

Corporeal Politics

Since 1959 Cuba's socialist health ideology, in part predicated on the idea that health care is a basic human right, has been successful both at the level of ideology and in practice. The tangible results of this success were evident in the country's health profile, as reflected in key World Health Organization indicators such as the island's low infant mortality rate and longer life expectancies at birth, among other vital statistics. Such quantifiable proof bolstered Fidel Castro's claims in the 1980s that Cuba is a world medical power. In effect, Cuba's socialist revolutionary period, also known as the Período Revolucionario Socialista (1959-present), engages in a kind of corporeal politics that has effectively produced a new kind of medicalized subjectivity, one in which a prolific network of health professionals has encouraged the citizenry to become increasingly attuned to biomedical understandings of what constitutes bodily health and physical well-being (Brotherton 2012). One of the results of embracing this subjectivity has been the increasing reliance on biomedical intervention and innovation. Physicians and their patients, particularly those who are ill, have become much more invested in what DelVecchio Good (2007) calls "a politics of hope," whereby the power of biomedicine, infused with a millenarian quality, takes center stage as the primary therapeutic answer. This socialist health care doxa[7] has saturated people's everyday lives and mundane practices, producing state-fostered expectations and feelings of entitlement to a particular form of (bio)medical health care.

In 1991 the Cuban government declared that socialism was under siege and formally announced the beginning of the Período Especial en Tiempos de Paz ("Special Period in Time of Peace," hereafter, período especial). This represented a distinctive rupture in institutionalized forms of comprehensive care, signaling a widening gap between the material benefits of the state's provision of basic human needs and the actual lived experience of Cubans.

The logic of everyday life in post-Soviet Cuba was thus radically transformed under the rubric of "war time measures in times of peace." Operating in many ways as a "state of exception,"[8] government policies institutionalized corrective measures by creating new juridical-legislative policies (migratory laws, banking practices, employment categories, and access to basic needs and services, to name but a few), and refining older ones, as part and parcel of a general program of economic recovery and revival.[9] In 1991, Julio A. García, the former head of the Cuban Chamber of Commerce, described the Communist Party's logic behind these changes as follows: "We have to think like capitalists but continue being socialists" (cited in Eckstein 1994: 103). As was evident from García's statement, within a context of "crisis" (both of the state and indivdiual bodies), the island started charting a new course for the social, political, and economic survival of the country's socialist revolution.

Over the past decade scholars and political commentators have continued to debate whether the período especial, as a transitory phase, has officially ended in light of the country's improving economic indicators in the late 1990s. Yet the rush to demarcate a beginning and an end obscures the lasting affective and corporeal dimensions of how this period was imprinted on people's bodies — in particular, how it was embodied through physical and mental ailments, palpably and materially experienced through deep senses of loss, betrayal, disillusion, and longing. The redefining of the socialist state through the lens of crisis directly influences the multifaceted ways in which individual Cubans in Cuba construct narratives about bodily and psychological health through the vagaries of social, economic, and political change. It is within the interstices of discursive and material changes to everyday life that a new biopolitical project is at work. It seeks to divert the moral expectations, assumptions, and entitlements of the citizenry away from the "cradle to grave" social welfare the socialist state so painstakingly fostered for over three decades to be more in line with the forces of market capitalism. Whereas the state was once the arbiter of health policy, care, and knowledge, a role its citizens were inculcated to accept via la Revolución, it now relies on transformed bodily practices that match new health agendas and new capital mobility.

Consultorio Los Molinos

In the 1980s, MINSAP, instituting a measure considered to be the most ambitious phase of Cuba's national health care system, initiated a proposal to train family physicians to participate in an innovative primary health care program called El Programa del Médico y la Enfermera de la Familia (MEF), or the Family Doctor-and-Nurse Program. By the end of 1985 the MEF program had expanded to include 10,000 physicians, and by 2009 it included 34,261 physicians overseeing 100% of the population. Physician-and-nurse health teams are stationed in consultorios located throughout the country. Each clinic serves 120 families, approximately 600-700 persons, and is located in a designated health area (área de salud) that the physician-and-nurse team serves.

In organization, the MEF program calls for family physician-and-nurse teams to live and work on the city block or in the rural community in which they serve. Moreover, physician-and-nurse teams are stationed in every factory and school. The role of the new family physicians, according to the MEF work program issued by health officials, was to carry out clinical and social-epidemiological vigilance of the population, promote health, and prevent disease by working in tandem with the community (MINSAP 1988). According to the MINSAP, the design and structure of the MEF program allowed for greater accessibility to health care services and a closer relationship between health teams and their patients. This tactic, the MINSAP proposed, would allow health teams an opportunity to obtain a more intimate knowledge of their patients and family members, enabling family physicians to better comprehend their patients' psychological and physical problems and to provide immediate, continuous care. Cuban leaders asserted the MEF program was the ultimate achievement in the field of health care. It went beyond the World Health Organization's Alma Ata Declaration of 1978 on primary health care, and made significant headway on a holistic approach to health care that addresses biological factors in tandem with an individual's material and social environment, thereby providing a diagnosis on the social fabric that encompasses health and well-being.

In the spring of 2000, I first visited Dr. Luis Pérez, a family physician, at his MEF consultorio in a small subdivision (reparto) of Los Molinos, which is located in San Miguel del Padrón, a working-class municipality southeast of Havana. The first floor of Pérez's two-story house consisted of a simple multi-room complex with a small examining room complete with posters for health promotion campaigns, nutritional advice, and various health programs to be followed, all plastered on the painted but crumbling walls. The second floor of the house is the physician's apartment, a modest two-bedroom unit with little furniture, a borrowed black-and-white television set, a broken bed, a kerosene stove, and a small Russian refrigerator. Pérez apologized for the poor physical condition of his apartment and said that the state had promised to improve his living conditions. He rolled his eyes and added half-heartedly, "At least, that is what the state has promised me, but you know how things are in Cuba. Currently, I am not a priority."

Referring in general terms to the patients in his área de salud, Luís Pérez discussed the various cases he was currently treating. Home visits, he made clear, were the only way he could get an adequate diagnosis on the chronic and acute problems in each household. While shortages had affected his daily practice, Pérez emphasized that with certain health programs, such as those for infants and pregnant women, he took absolutely no risks. He would personally and immediately notify the local polyclinic or municipal hospitals of any shortages or required medicines or supplies. The death of an infant in one's área de salud, he cautioned, was subject to serious investigations by the municipal authorities. Pérez had to meet once a month with a clinical supervisor, an obstetrician-gynecologist from the policlínico (a municipal-based clinic with more specialists present to serve a larger patient population), to review all of his pediatric cases and prenatal appointments. The maternal and infant care programs were the most important of his daily activities. He spent hours reviewing his notes and getting them in order before the supervising obstetrician-gynecologist made her monthly rounds in his área de salud. Not having a computer, Pérez assembled the majority of his files and notes in a binder. For the most part, as stipulated by the MEF program, adult patients kept their own clinical history as well as that of their underage children and were required to bring their file with them on visits to the family physician, policlínico, or local hospitals.

When I returned in 2004, Consultorio Los Molinos was vacant. Louis Pérez had been selected to join the MBA in Venezuela; after a year there he returned to Cuba for what became a short-lived stay. Within several weeks of his return he was once again dispatched with a medical brigade team, this time to Pakistan after the earthquake in October 2005. In his absence, the conditions of the neighborhood he serviced in Los Molinos deteriorated significantly. Many of the patients who had regarded Pérez as their first line of defense for many of their medical emergencies had taken to self-medicating and making ad hoc diagnoses spread by word of mouth.

Patients that I had interviewed prior Pérez's departure, such as Bení Ortiz, with whom the physician was personally working to address his alcoholism, was consumed by the illness when I visited in 2005. Since the consultorio was closed, Mariella, Bení's wife, was forced to visit the municipal policlínico to see if she could get her husband into some kind of treatment program. The inconvenience of getting there, compounded by the lack of personal care by the attending physicians, quickly led Mariella to abandon the prospect of returning to the policlínico. Tired of the drama, as she called it, she tried to make her problems more bearable by indulging in a steady diet of coffee and sleeping pills – a problem that Dr. Pérez was equally aware of and had worked tirelessly to treat when he was there. Mariella asked, "Had all of Pérez's work been in vain?"

In a reversal of roles, Mariella's concerns echoed those of people in countries that had been the recipients of provisional humanitarian medical aid: what happens when the doctors leave?[10] In an ironic twist, Pérez's departure to engage in humanitarian work in other parts of the globe left his own patients without the personalized care that had been the very benchmark of the MEF program. Data collected from Cuba's Office of National Statistics reflects the changing practice of health care by reporting a 22% increase in the number of emergency visits to policlinics between 2004 and 2006.[11] Because most of the patients in his ficha familiar[12] were categorized, classified, and tabulated for individualized health care programs under the dispensarización program, Pérez's departure left a void.[13] The overworked staff at the local policlínico had neither the time nor the mandate to circulate in the communities they served. As Mariella complained, the health workers were, in effect, strangers; they quickly reviewed your file, wrote prescriptions, and ushered you out the door. They could not comment on or assess your living conditions, whether you had enough to eat, or even if you had a roof over your head. This signaled the return to an impersonal, disembodied style of health care delivery. Many of the residents in Pérez's área de salud nostalgically recalled the days when having a doctor on the block or en el terreno (in the field) making house calls was a luxury. In effect, they had lost their personalized "first line of defense."

Shortly after he returned to Cuba in 2006, I interviewed Pérez. He was preparing to depart for Venezuela on another mission. The formerly tall, lanky physician was now considerably heavier. "Viste! Estoy gordo!" (Look, I am fat!), he proudly exclaimed when we met at my apartment. Riffling through the pictures he had taken in Pakistan and Venezuela, Pérez excitedly described the travails of his life as an internacionalista. While the hardships of life in Venezuela were trying, he enjoyed his work. His new patients, he said, were extremely appreciative and warm. He described the extreme pride he felt in finally being more financially secure and able to realize many of his dreams of traveling and having certain material luxuries. During his absence, the state continued to deposit his monthly peso salary (about 20 U.S. dollars) into a bank account in Cuba as well as a monthly stimulus of 50 pesos convertibles (52 U.S. dollars). The stimulus funds could be accessed only upon his return to the island. These payments, he added, were in addition to the stipend of approximately 180 U.S. dollars per month he received in Venezuela. Financially, he was in a privileged position now.

Pérez had arrived on the island with several cargo boxes of luxury goods, including a television set, a DVD player, and a stereo, brand-name clothes, and some small pieces of furniture. Several of his patients were vocal in their complaints about the physician's newly acquired wealth. Another family doctor in a nearby community who was required to take on Pérez's caseload of infant and maternal health cases, in addition to her already busy caseload, was boiling when I interviewed her: "Nobody pays me a higher wage for doing twice the amount of work I used to. We all have to make sacrifices for this profession, so it really annoys me to see these internacionalistas talking about humanidad [humanity]. It is also a shopping expedition. You have to be chosen to be an internacionalista. Trust me, it is like finding a pot of gold to be able to go, especially in the crisis we are experiencing now."

Aware of what he termed "an epidemic of envidia [envy]" Pérez had made a concerted effort to bring regalitos (gifts) for many of his patients and colleagues to defuse the issues of his long absence and the symbols of his new jet-setter lifestyle. This was all the more jarring given that many of the people in his community did not even have a passport, much less had ever boarded a plane in their lives. An elderly woman I interviewed in Pérez's área de salud, commenting outspokenly on the Venezuelan oil-for-aid deal, remarked bitterly, "My doctor is helping to keep the lights on" (prior to Chávez's agreement, apagones (blackouts) were a regular occurrence throughout Havana). Her sentiments were reflected in other interviews I conducted, in which citizens increasingly vocalized their anger and resentment at the loss of medical personnel as a result of what several individuals referred to as "their sale" to Venezuela. Peoples' conversations suddenly took on a more critical tone. During my visit to the island in 2005, the sudden appearance at the local bodega and dollar stores of sundry items marked "Products of the Bolivarian Revolution, Venezuela," were met with great fanfare and excitement. The canned sardines, packages of chocolate drink mix, and tinned meats at reasonable prices were welcome additions to the limited fare traditionally available. When I returned in 2007, these very same items were met with derision and disappointment. Several people asked, "What exactly were the benefits of the agreement with Venezuela for the average citizen?"

Surplus and Deficits in a Shifting Moral Economy

The concerns that citizens raised were, in many ways, prompted by several physicians' confessions that they were eager to be selected as internaciona-listas. For a profession that is typically remunerated very little in comparison to private business or tourism, many physicians highlighted their view that material rewards were not insignificant. "I can't live off of feeling good," one physician joked when I prodded him that he was, theoretically, participating in these missions to do good. His response revealed that the humanitarian ethic was merged with a personal desire not only to travel, but also to acquire material goods.

When I interviewed Dr. Alberto Navarro in 2007, he had just returned from Caracas, where he was working as a consultant with the Venezuelan Ministry of Public Health. He was unhappy about his temporary return to Havana. "The city looks so dark," he complained. The main thoroughfares in Havana were dimly lit. The streetlights either had blown out and not been replaced or were broken. He had returned to Cuba only to visit his family and purchase a house in Cienfuegos, an incentive the government provided for some of the returning internacionalistas.[14] This created some tensions in his previous área de salud.

While his old consultorio stood vacant, Navarro was now in the pro-cess of setting up house a couple of blocks away, complete with the material luxuries he had acquired abroad. To make matters worse, he announced that he would be extending his contract for an additional two years in Venezuela. Planning to leave his house with family members, he intended to "make the best" of his time in Venezuela. He was well aware that several of the residents in his área de salud had responded to this announcement by calling him a medical jinetero, hustling medicine for money. He had, in their eyes, reduced himself to a commodity. Hurt by these comments, he said that while these trips made good financial sense, Caracas was not an easy place to live. The constant fear of crime and the resentment of various sectors of Venezuelan society, including the middle class and the elite, on the growing presence of Cubans in their country made life there tense. As he concluded, ultimately, he had to live in a foreign country without seeing his family for long stretches of time. This was a sacrifice he was willing to undertake to be able to save enough money to return to Cuba and live well. This was not, he added, an unworthy or unjust pursuit.

While Cuba boasts of one of the highest physician-per-inhabitant ratios in the world, the massive deployment of doctors on foreign missions has left noticeable absences in the country's domestic primary health care programs. In his address to the nation on 26 July 2007, then-acting president Raúl Castro called on Cubans to hold meetings to discuss the country's most pressing problems. In the public community meetings that ensued, citizens became increasingly vocal about the daily constraints they encountered in trying to solve their immediate health problems. In March 2008 Raúl Castro announced that the MEF program would be reorganized to address some of the structural and staffing problems that were created as a result of the increasing foreign demands for the country's primary health care physicians. Rather than emphasize consultorios, the capstone of Cuba's primary health sector, polyclinics were to take on a considerably more important role. The aim of these changes, health officials noted, was to strengthen their roles in the communities they serviced and to add a host of specialty services previously only available in hospitals, including X-rays, ultrasounds, psychiatry, and cardiology, among other services.

Most individuals were content with the added services. However, many were equally saddened by the slow disappearance of the family physicians stationed in their respective communities. The traditional physician en el terreno, or making house calls and visiting patients with chronic health problems, was increasingly becoming the exception rather than the rule. As many citizens point out, the Cuban model of primary health care being exported is, in many ways, not the one being practiced at home. While individual complaints have not translated into poor health outcomes (for example, Cubans are not dying for lack of a physician), and the country's health profile remains stable amidst massive socioeconomic changes, the popular support for medical internationalism such as MBA has decreased among certain sectors of the population. For example, one individual commented, "Sometimes I wonder if we weren't better off with the lights off. At least we weren't forced to see what we were lacking."

At a May 2009 conference focused on assessing the Cuban revolution on its fiftieth anniversary, there was a special panel devoted to Cuba's medical internationalism. After the various presentations, audience members asked about complaints that the medical needs of Cubans in Cuba were no longer being met because of MBA and other medical missions. Various panelists quickly dismissed them. As one panelist stated, such complaints needed to be put in the context of "an overmedicalized population, who have been spoiled by too many years of having a doctor on every block." The conclusion to

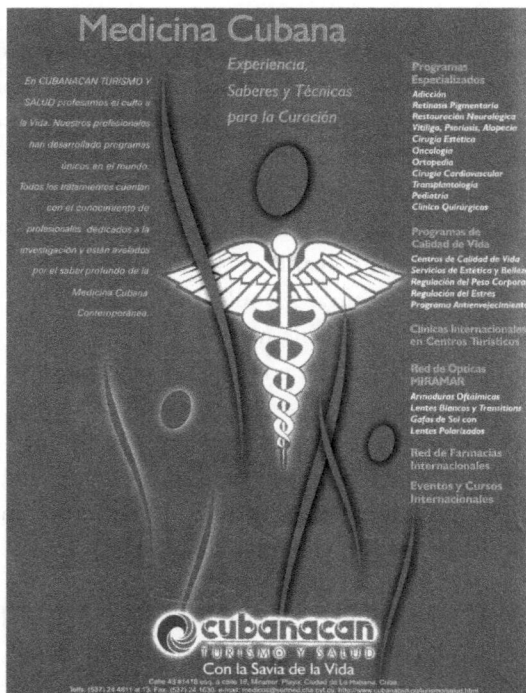

Figure 5.3: Cubanacán Tourism and Health.
Source: Avances Médicos de Cuba (2000) 7.23, back cover

be drawn here: Cuba does not need so many physicians. While there may be much truth to this conclusion, it does not take away from the lived experience of Cubans who, for more than 50 years, have been trained by the state not only to accept, seek out, and desire medical intervention – from the most basic first-aid problem to the most severe medical matter – but, equally, to feel entitled to and expect those very services. The Cuban Revolution built its legitimacy, in part, on the provision of a universal, accessible health care system as a basic human right.

Cuban health officials increasingly assert that it is the very success of the primary health programs of the past, such as the MEF, that have made it possible for Cuba to begin the process of restructuring health care to meet the populations' new health needs. Nevertheless, another narrative emerges from the perspective of the average citizen. That is, the *moral economy of the gift, as creating bonds of solidarity, is increasingly being called into question*. While daily shortages during Cuba's socialist and economic crisis have become embodied in people's everyday lived experiences, the sudden deficit of physicians is created by entirely different market rationalities. Physicians, traditionally understood to possess no economic value, but who impart their knowledge and expertise through a moral imperative of socialism, are now entangled in an economy of exchange. The surplus of Cuban-trained physicians who were supposed to cure the social ills of society and work to foster socialist morals and values are now luxury commodities, like those advertised in Cuba's health tourism campaigns (known as Turismo y Salud; see Figure 5.3), bartered and contracted out to foreign destinations to serve as accessible, affordable medical labor.[15] This characterization of the revolutionary physician as a commodity for export speaks to another kind of a moral economy of exchange that warrants further consideration.

Transactional Humanitarianism in the Post-Soviet Era

In the late 1960s and particularly the early 1970s, official state discourse in Cuba was steeped in Marxist-Leninist and Guevarist principles. Fidel Castro advocated that under communism people would contribute according to their capacity and be rewarded according to their need. Rather than work for material rewards, workers were to labor out of a sense of moral commitment and to be recognized with pennants, flags, and titles. This constituted a specific kind of "moral economy," a term elaborated upon by James Scott (1976), whereby egalitarian forms of reciprocal exchange took center stage over economic gain, profit, and "the market."[16] For instance, during the Year of Solidarity in 1966, Fidel Castro (1969: 199) stated, "If we want people to remove the dollar sign from their minds and from their hearts, we must have men who have gotten rid of their own mental dollar signs." In line with these statements, the government nationalized and expanded social services. Education, medical care, social security, day care, and most housing was provided free of charge, with access to them designed to be more equitable and need-based than ever before. This helped bolster the government's rhetorical claims by putting them into practice.

Cuban sociologist and historian Haroldo Dilla (2001) argues that prior to 1989, 94% of the workforce in Cuba was employed in state enterprises; workers were divided into about 20 salary categories with fixed remuneration matched by subsidized consumer goods. However, by 1996, during the período especial, the percentage had shrunk to 78, and a significant portion of the population had moved into the private, mixed, and cooperative sectors. Relatively high amounts of market-based wealth and power began to concentrate in the hands of a small group of people, especially in economically expanding regions such as those catering to tourists. Cuba's economy of remittances also contributed notably to the growing income disparity among groups in the population. Official figures estimate that remittances increased from a reported 50 million dollars in 1990 to over 900 million in 2005, representing a large infusion of foreign currency circulating in the hands of individual Cubans during the this period.[17]

Some of the consequences of the changing nature of Cuban socialism in the post-Soviet context have been the creation of a situation wherein Cubans without regular access to foreign currency now look with envy at the minority, who can enjoy the fruits, sometimes quite conspicuously, of the two-tiered economy (Brotherton 2012). Many of these "conspicuous consumers," in times of apparent scarcity, now include physicians returning from medical missions. This contributes to the fact that the moral character of past state campaigns, where the "gift," as embodied in physicians' labor, was imagined to circulate as a means to create social bonds,[18] now carries considerably less material and rhetorical authority. This is particularly salient, as individual citizens must negotiate the tenuous, sometimes ill-defined withdrawal of the state in the social and material well-being of their everyday lives. This changing political economy has produced a decidedly different kind of moral universe, one in which "the market" is a lived reality to be confronted on a daily basis rather than an abstraction or sublimated phenomena.

The Marxist framework of "accumulation through dispossession," as articulated by David Harvey (2003), offers prodigious insight for understanding the economy of surplus and deficits emerging in Cuba's post-Soviet era (see, for example, Brotherton 2008). By the same token, the "gift," in a context of scarcity and excess, is also reminiscent of French intellectual Georges Bataille's (1985) notion of expenditure, or dépense, where great amounts of loss are tied to ostentatious displays of wealth. The analytical thrust of this chapter, therefore, is not to argue that gifts and commodities are mutually exclusive categories. Rather, I argue that similar to the "lived experiences of socialism," many of the emerging strategic medical missions are paradoxical. Such missions must, now more than ever before, skillfully vacillate and/or blur the distinction between savvy socialist corporations and socialist humanitarians. While seemingly oppositional, both characterizations speak to what I argue is the Janus-faced nature of medical internationalism. We must highlight the "transactional" nature of strategic forms of humanitarian aid, such as Barrio Adentro, in order to trouble the idea of the "gift" as strictly a benevolent act that forges reciprocal bonds. Many everyday citizens have already noted, and similarly draw attention to, the messy interplay involved in the exchange or transfer, sometimes skewed or unequal, of goods, services, or funds. It is also just as important to view such transactions as deeply social events, involving multiple actors or things that affect or influence each other. In this respect, the impact of medical missions on Cubans in Cuba should be not be silenced by the cacophony of voices, both on and off the island, of countless officials, analysts, scholars, or activists who are entangled in a kind of corporeal politics of social justice.

Unlike the mid-1960s and 1970s when Cuba, as part of its program to "export Revolution," was active in insurrection movements throughout the Americas (and to a lesser extent in Africa), the government's foreign policy has now, more explicitly, traded guns for stethoscopes. In this way, my focus on transactional humanitarianism is not meant to strictly point out the obvious in Cuba's post-Soviet era: that market rationalities are increasingly driving the exchange of physicians across the globe. The marriage of humanitarianism and explicit or implicit (or both) economic and/or political agendas are not unique or even limited to Cuba's medical internationalism programs. The commodification of humanitarianism is not a radical departure from Cuba's foreign aid policies of the past.[19] However, it is my assertion that this very circulation of physicians, more appropriately understood as a form of "friction" (Tsing 2005), must happen through moral valences that lay bare the myriad ways in which the specter of the Soviet past and the uncertainty of the island's political future have served as potent signifiers of the nation's vulnerability, particularly as the withdrawal of Soviet aid and the magnified effects of the U.S. embargo manifest at the level of individual bodies and reverberate through the multiple spheres of quotidian life. From a location of "exclusion" and/or "exception," Cuba's increased focus on transactional humanitarianism as a strategy to, literally and metaphorically, fuel la Revolución should serve as a much-needed case study for helping us refine our current – and indeed limited – understandings of the state, statecraft, and subject formation in varying socioeconomic contexts.

148   P. Sean Brotherton

References

Agamben, G. (2005) *State of Exception*, Chicago, IL: University of Chicago Press.
Andaya, E. (2009) 'The gift of health: socialist medical practice and shifting material and moral economies in post-Soviet Cuba', *Medical Anthropology Quarterly*, 23.4: 375–74.
Bataille, G. (1985) *Visions of Excess: Selected Writings, 1927-1939*, ed. and trans. Alan Stoekl, Minneapolis: University of Minnesota Press.
Bornestein, E. and Redfield, P. (2010) *Forces of Compassion: Humanitarianism Between Ethics and Politics*, New Mexico: School for Advanced Research Press.
Briggs, C. and Mantini-Briggs, C. (2009) 'Confronting health disparities: Latin American social medicine in Venezuela', *American Journal of Public Health*, 99.3: 549–55.
Brotherton, P.S. (2008) '"We have to think like capitalist but continue being socialists": medicalized subjectivities, emergent capital, and socialist entrepreneurs in post-Soviet Cuba', *American Ethnologist*, 35.2: 259-274.
— (2011) 'Health and health care in Cuba: history after the revolution: key phases and overviews of health development', in A. West-Dúran (ed.) *Cuba: People, Culture, and History*, New York: Charles Scribner's Sons.
— (2012) *Revolutionary Medicine: Health and the Body in Post-Soviet Cuba*, Durham, NC: Duke University Press.
Brouwer, S. (2011) *Revolutionary Doctors: How Venezuela and Cuba are Changing the World's Conception of Health Care*, New York: Monthly Review Press.
DelVecchio Good, M.J. (2007) 'The medical imaginary and the biotechnical embrace', in J. Biehl and A. Kleinman (eds.) *Subjectivity: Ethnographic Investigations*, Berkeley, CA: University of California Press.
Dilla, H. (2001) 'Local government and economic and social change in Cuba', *FOCAL: Canadian Foundation for the Americas*, (May): 1–7.
Eckstein, S.E. (1994) *Back from the Future: Cuba under Castro*, Princeton: Princeton University Press.
— (2009) *The Immigrant Divide: How Cuban Americans Changed the U.S. and Their Homeland*, New York: Routledge.
Economist Intelligence Unit (E.I.U.) (1999) Cuba: A Country Profile.
Fassin, D. (2007) 'Humanitarianism as a politics of life,' *Public Culture*, 19.3: 499-520.
— (2010) 'Noli me tangere: the moral untouchability of humanitarianism', in

D. Fassin and M. Pandolfi, *Contemporary States of Emergency: The Politics of Military and Humanitarian Interventions*, New York: Zone Books.

— (2011) *Humanitarian Reason: A Moral History of the Present*, Berkeley, CA: University of California Press.

Fassin, D. and Pandolfi, M. (2010) *Contemporary States of Emergency: The Politics of Military and Humanitarian Interventions*, New York: Zone Books.

Feinsilver, J.M. (1993) *Healing the Masses: Cuban Health Politics at Home and Abroad*, Berkeley, CA: University of California Press.

— (2008) 'Oil for doctors: Cuban medical diplomacy gets a little help from a Venezuelan friend', Nueva Sociedad 216. Online. Available HTTP: <http://www.nuso.org/upload/articulos/3537_2.pdf>(accessed 12 July 2011).

Feldman, I. and Ticktin, M. (2010) *In the Name of Humanity: The Government of Threat and Care*, Durham, NC: Duke University Press.

Fox, R. (1995) 'Medical humanitarianism and human rights: reflections on Doctors without Borders and Doctors of the World', *Social Science & Medicine*, 41.12: 1607–16.

Guevara, E. (1968) 'On Revolutionary Medicine', in J. Gerassi (ed.) *Venceremos: The Speeches and Writings of Che Guevara*, New York: Macmillan.

Harvey, D. (2003) *The New Imperialism*, Oxford: Oxford University Press.

Huish, R. (2008) 'Going where no doctor has gone before: the role of Cuba's Latin American School of Medicine in meeting the needs of some of the world's most vulnerable populations', *Public Health*, 122: 552–57.

— (2009) 'How Cuba's Latin American School of Medicine challenges the ethics of physician migration', *Social Science & Medicine*, 69.3: 301-304.

Huish, R. and Kirk, J.M. (2007) 'Cuban medical internationalism and the development of the Latin American School of Medicine', *Latin American Perspectives*, 34.6: 77–92.

Huish, R. and Spiegel, J. (2008) 'Integrating health and human security into foreign policy: Cuba's surprising success', *International Journal of Cuban Studies*, 1.1: 1–13.

Kirk, J.M. (2009) 'Cuba's medical internationalism: development and rationale', *Bulletin of Latin American Research*, 28.4: 497–511.

Kirk, J.M. and Erisman, M. (2009) *Cuban Medical Internationalism: Origins, Evolution, and Goals*, New York: Palgrave Macmillan.

Lakoff, A. (2010) 'Two Regimes of global health', *Humanity: An International Journal of Human Rights, Humanitarianism, and Development*, 1.1: 59–79.

Mauss, M. (1990) *The Gift: Forms and Functions of Exchange in Archaic Societies*,

London: Routledge.

Ministerio de Salud Pública (MINSAP), República de Cuba (1988) *Médico de la Familia: Información Estadística*, Havana.

— (2001) *Carpeta Metodológica de Atención Primaria de Salud y Medicina Familiar*, VII Reunión Metodológica del MINSAP, Havana.

Nguyen, V. (2009) 'Government-by-exception: enrolment and experimentality in mass HIV treatment programmes in Africa', *Social Theory and Health*, 7: 196–217.

— (2010) *The Republic of Therapy: Triage and Sovereignty in West Africa's Time of AIDS*, Durham, NC: Duke University Press.

Ong, A. (2006) *Neoliberalism as Exception: Mutations in Citizenship and Sovereignty*, Durham, NC: Duke University Press.

Pandolfi, M. (2003) 'Contract of mutual (in)difference: governance and humanitarian apparatus in Albania and Kosovo', *Indiana Journal of Global Legal Studies*, 10.1: 369–81.

— (2007) 'Laboratory of intervention: the humanitarian governance of the post-Communist Balkan Territories', in S.T. Hyde, B. Good and S. Pinto (eds.) *Postcolonial Disorders*, Berkeley, CA: University of California Press.

Pérez, O. and Haddad, A.T. (2008) 'Cuba's new export commodity: a framework', in M. Font (ed.) *Changing Cuba/Changing World*, New York: Bildner Center for Western Hemisphere Studies, CUNY.

Peterson, K. (2012) 'AIDS policies for markets and warriors: dispossession, capital, and pharmaceuticals in Nigeria', in K.S. Rajan (ed.) *Lively Capital: Biotechnologies, Ethics and Governance in Global Markets*, Durham, NC: Duke University Press.

Polanyi, K. (1957) *The Great Transformation*, Boston, MA: Beacon.

Redfield, P. (2005) 'Doctors, borders and life in crisis', *Cultural Anthropology*, 20.3: 328–61.

— (2006) 'A less modest witness: collective advocacy and motivated truth in a medical humanitarian movement', *American Ethnologist*, 33.1: 3–26.

Rojas, M. (1986) *El Médico de la Familia en la Sierra Maestra*, Havana: Editorial Ciencias Médicas.

Scott. J. (1976) *The Moral Economy of the Peasant*, New Haven, CT: Yale University Press.

Tsing, A. (2005) *Friction: An Ethnography of Global Connection*, New Jersey: Princeton University Press.

Thompson, E.P. (1971) 'The moral economy of the English crowd', *Past and Present*, 50: 76-136.

Endnotes

1   For instance, Marta Rojas (1986), a widely read Cuban journalist, wrote a book entitled Doctors in the Sierra Maestra, which chronicles the relentless sacrifice of physicians in the rural Sierra Maestra mountain ranges, a geographic region symbolically tied to the origins of the revolutionary guerrilla movement.

2   See the works of Huish and Kirk 2007; Kirk and Erisman 2009; Kirk 2009.

3   See, for example, Borstein and Redfield 2010; Feldman and Ticktin 2010; Fassin 2007; Fassin 2011; Fassin and Pandolfi 2010; Fox 1995; Nguyen 2009, 2010; Pandolfi 2003, 2007; Peterson 2012; Redfield 2005, 2006.

4   Most contemporary humanitarian efforts in the field of global health address what anthropologist Andrew Lakoff (2010) identifies as two regimes of intervention: global health security and humanitarian biomedicine. While not mutually exclusive, global heath security, he argues, is more concerned with global disease surveillance and targeting national public health infrastructure. Humanitarian biomedicine (e.g., MSF), on the other hand, is concerned with addressing the lack of adequate access to basic health care needs.

5   The first agreement, known as the "Convenio Integral de Cooperación entre la República de Cuba y la República Bolivariana de Venezuela" (Integral Cooperation Accord), was signed on 30 October 2000. In April 2005 Castro and Chávez signed the Regional Integration Project to expand, build, and collaborate on various joint projects, including the MBA program. See also Briggs and Mantini-Briggs (2009) for a discussion of the program.

6   As Feinsilver notes (in this volume), earnings from medical services reportedly equaled 28% of the total export receipts and net capital payments in 2006. This amounted to 2,312 million U.S. dollars, generating more money than both nickel and cobalt exports and tourism.

7   Bourdieu (1977), very broadly, describes doxa as the taken-for-granted assumptions that individuals hold as being self-evident and unquestioned truths in society.

8   Within times of crisis, Agamben (2005: 5) asserts, the "state of exception" refers to the expansion of the powers of government to issue decrees that have the force of law. In the process of claiming this power, questions of sovereignty, citizenship, and individual rights can be diminished, superseded, and rejected.

9   For instance, the government introduced reforms that sought to restore import capacity and stimulate domestic supply; increase the economy's responsiveness to the world market; search for foreign capital and technology; allow free-

market sales of surplus produce, handcrafts, and some manufactured goods; increase the categories of self-employment allowed by the state to cover an additional 100 freelance occupations; and permit the registration and taxation of private rental activity (E.I.U. 1997).

10   See Redfield's work on Médecins Sans Frontières [MSF] (2005, 2006).

11   An analysis of these figures can be found in Pérez and Haddad (2008).

12   This is a record of preventive services and conditions for all patients in their area. This record is updated and reviewed at least monthly with a clinical supervisor who is an academically based family physician. Acute and chronic health problems are coordinated in a database at municipal, provincial, and national levels of the MINSAP. The monitored services and conditions include prenatal and natal care, immunizations, cancer screening by smear and mammography, risk factors such as smoking, hypertension, and follow-up for chronic conditions as well as psychosocial problems and sources of stress in the family or at work.

13   Under the patient classification surveillance system of dispensarización, the physician-and-nurse health teams evaluate the health situation in a specific area and define the at-risk populations by patient, for example: hypertensive, diabetic, expectant mother, and so on. Each patient then receives an assigned priority and differentiated treatment in accordance with nationally prescribed procedures and programs appropriate for their age, gender, and risk factors. See Brotherton (2011, 2012) for more extensive discussion of the specific objectives of the MEF program.

14   In August 2007, U.S. officials announced the Cuban Medical Professional Parole program that allows Cuban medical personnel, identified by the Department of Homeland Security as doctors, physical therapists, laboratory technicians, nurses, sports trainers, and others, to apply for entry into the United States at U.S. embassies in the countries where they serve. In essence, they opened the door for Cuban personnel on missions to defect to the United States. Cuba's generous incentives to return to the island could be a response to this policy.

15   See Brotherton (2008) for an examination of the state's promotion of health tourism in Cuba.

16   See, for example, the important works of Polyani 1967; Mauss 1990; and Thompson 1971 on defining how the moral economy operates vis-à-vis market economies.

17   There is a growing body of literature that addresses the critical role of remittances in both the formal and informal economy in Cuba (see, for example, Eckstein 2009).

18 See Mauss's (1990) anthropological discussion of the role of the gift in establishing reciprocal exchanges; see also Andaya (2009) on the "gift of health" in
Cuba's post-Soviet medical sector.

19 Throughout the 1970s and early 1980s, Cuba, with the financial backing of
the Soviet Union, was in a privileged position to export and also capitalize on
marginalized but wealthier nations' desires to replicate Cuba's public health
experience. While in the 1960s Cuba's focus was on humanitarian medical
missions, this changed in the 1970s, when it began to charge oil-rich nations
such as Libya and Iraq, which paid in hard currency, on an ability-to-pay basis.
This became the foundation for early trade agreements (see, for example,
Eckstein 1994).

6

Transformative Medical Education and the Making of New Clinical Subjectivities through Cuban-Bolivian Medical Diplomacy

Alissa Bernstein

Introduction: Celebrating Medical Diplomacy

On a sweltering humid afternoon in August of 2009, I took a crowded bus to a beach town outside of Havana to attend a medical school graduation party for three Bolivian students who had just completed their medical studies in Cuba. These students studied for six years in Cuba affiliated with la *Escuela Latinoamericana de Medicina*, the Latin American School of Medicine (ELAM), where they spent time in practical training in clinics throughout the country. The school is part of a larger apparatus of medical diplomacy, wherein Cuban medical expertise and infrastructure has, in recent years, been a key means through which the Cuban state builds relationships with other countries, mostly in the Global South (Feinsilver, this volume). At the party, the medical students held the center of attention, photographed with their diplomas and congratulated by all who attended, as the festivities slowly turned into a dance party on the tiny balcony of the two-story house once the sun began to set.

As we ate plates of the standard Cuban fare of rice, beans, and chicken, I spoke in a quieter corner with Ernesto,[1] a medical student in his early twenties from Bolivia. Ernesto was studying in a practice-based program in Cuba's Cienfuegos province in a rural polyclinic. During his medical training, Ernesto worked closely and lived with family doctors in order to learn not only the medical curriculum, but also medicine as a vocation, including the "way of life" of a doctor as a community leader and educator. He told me:

Living in Cuban society with families and studying with Cuban doctors you learn so much about the ideas behind general health care. We learn about the system by seeing how people live in Cuba and drawing our own conclusions. Because of this there is a big difference today in who I am. I am not the same person who came here four years ago. My mind has changed and I have learned so many things that will be useful in my practice and for my country.

"What do you mean 'your mind has changed'?" I asked. Ernesto explained:

Before I came here I was selfish and wanted to be the best. Here I have learned many principles about the ethics of our profession: to be honest, kind, to treat people well, to work on relations with patients. With what I have learned I want to try to make changes in my own country where there is now a new government led by Evo Morales that is trying to create a new health care system. I feel that many changes are now possible.

In this discussion, Ernesto made a discursive move from personal and professional growth to national engagement. He expressed hopes that his return and medical practice could have the potential to aid in the larger health system transformations occurring in Bolivia. When I asked him how he might do this, he was full of ideas. He explained his desire to start a project to create a collaborative group of doctors to work with communities and families to teach people how to manage illnesses through preventive care. He also expressed his interest in teaching in a medical school in order to show other students what he learned in Cuba. The trajectory of his personal transformation through medical education abroad also provided the arc for the transformations that he envisioned for his country.

Two years later, inspired by these Bolivian medical school graduates and hoping to follow their story to Bolivia in order to explore a new angle in the Cuban medical diplomacy engagement, I attended another celebration, this time in El Alto, Bolivia. El Alto is known as the "Aymara capital of the world" because it is home to the largest population of indigenous Aymara people in Bolivia. The city sits at nearly 13,000 feet above sea level on the mountains overlooking La Paz, and is the site of a number of Cuban medical aid programs, such as *Operación Milagro* (Operation Miracle), the renowned Cuban ophthalmological program, as well as Cuban-run hospitals that provide free medical care to Bolivian residents (Albro 2005). On a sunny afternoon, I sat amongst a crowd of doctors, medical students, diplomats, former patients, and local politicians at the five-year anniversary celebration of Hospital Francisco Amistad Cuba-Bolivia,[2] which is run by Cuban doctors. The celebration for Hospital Francisco began with the singing of the Bolivian and Cuban anthems. Speeches that followed were interspersed with shouts of *"Viva Bolivia! Viva Cuba!"* and reference to the "vision of exchange" between the two nations. Indeed, Hospital Francisco is home to a new development in Cuban medical diplomacy efforts abroad. In 2010 the Cuban government implemented a new dimension of the Cuban medical education program here. Bolivian medical students study medicine in Cuba through their fourth year and then return to this hospital to finish their fifth and sixth years of medical training, where they do rotations in medical specialties such as urology, psychiatry, and surgery.

During an interview, Dr. Gonzalez, the sub-director of Hospital Francisco, explained the reasoning behind this change to the standard Cuban program, which normally has students spend their entire training in Cuba:

> Students return to Bolivia to finish their studies so that they can learn about how to relate with Bolivian patients and the pathologies that exist here, in their own environment. Diseases are different here [in Bolivia], and certain sicknesses are much more frequent. For example, in Cuba we don't have things like tuberculosis or Chagas, but the Bolivian students need experience with these in order to do good work in Bolivia.

Hospital Francisco is thus understood as a site where Cuban-trained Bolivian medical students can learn to mediate the boundaries between their training under the Cuban model of health care and the realities they will face as doctors in Bolivia. They are offered what is framed as an essentially Cuban form of medical expertise, to be used in what has been reframed as a Bolivian-specific medical context for treatment. Interviews with seven of the Bolivian medical students studying at Hospital Francisco revealed commonalities with Ernesto in Cuba: both their experiences of personal transformation as medical practitioners through their studies in Cuba and under the instruction of Cubans in Bolivia, and their desires to help transform their nation through their medical work in Bolivia.

The health practitioners and diplomats I interviewed, both Cuban and Bolivian, often essentialized the training and skills gained in Cuba, and the health contexts through which they traveled, as falling within the bounded national and cultural entities of "Cuba" and "Bolivia." In this chapter, I focus on the ways Cuba and Bolivia became framed as culturally and medically bounded sites through the stories of personal transformation the health practitioners told, what I call "transformation narratives," and through discursive contrasts they made between Cuba and Bolivia through the lens of their medical education in Cuba. I examine these framings to understand the ideas about health care and medical practice that were made mobile as a result and how they were put into circulation. In particular, I found that for the health practitioners I spoke with, their training made possible new medical imaginations in Bolivia that they understood as "Cuban," specifically in regards to the ways they interacted with their patients and their conceptions of community-based medicine. They felt they could package aspects of their training to implement in Bolivia. In following these visions, they often created spaces for community work where it did not already exist in current medical practice in Bolivia, such as mobile clinics, health outreach programs, and providing free health care to patients. This packaging and mobility of what they considered to be a "Cuban" approach to health care provision thus influenced the ways they participated in the health system in Bolivia as Bolivian citizens.

I build on work in medical anthropology on medical education and the development of "clinical subjectivity" (Good 1994; Stonington 2011) in conjunction with work on global health (Adams et al. 2008; Clarke et al. 2003) and medical humanitarianism (Redfield 2005, 2008; Fassin 2007). I look at medical education as not just the formation of a professional subject, but also a catalyst for personal growth and a space of formation that can reshape ideas about national participation and nation building as part of a transnational medical diplomacy engagement. I am interested in exploring the complexities of practices and discourses that circulated as these practitioners - and their skills and worldviews - developed, moved, and traveled into spaces of medical practice in Bolivia.

Research Methods and Ethnographic Settings

This chapter is based on one month of ethnographic fieldwork in Cuba in the summer of 2009 and five non-consecutive months of ethnographic fieldwork in Bolivia between 2010 and 2011. This research was approved by the Institutional Review Board for Human Subjects from the University of California, Berkeley. I also received official permission to conduct this research from the Cuban Embassy in Bolivia, which allowed me to interview Cuban doctors as well as Bolivian medical students and to do participant-observation in polyclinics, hospitals, and medical training programs run by Cubans in Bolivia.

Methodologically, I used semi-structured ethnographic interviews and partici-pant-observation, focusing on two groups of Bolivians involved in the health sector who have also studied in Cuba. First, I interviewed Bolivian medical students in Cuba who studied at ELAM and Bolivian medical students in Bo-livia who studied at ELAM for four years and were completing their last two years of medical training in Hospital Francisco in Bolivia under the direction of Cuban medical doctors. Second, I interviewed and followed Cuban-trained Bolivians who had completed their studies in Cuba at the *Facultad de Medicina Moron-Ciego de Avilain* in the areas of epidemiology, optometry, physiotherapy, and as health technicians, and who were working as health practitioners in Bo-livia. Much of the work on Cuban medical diplomacy currently focuses on the Cuban doctors or international students trained at ELAM. However, I chose to focus on medical students, health technicians, and public health specialists as a way to emphasize how medical care is often structured in Cuba, where physicians, social workers, and medical technicians work together through a team-based approach within the neighborhood (Feinsilver 1993). This pro-vides a way to expand the concept of clinical subjectivity as not just the gaze of the primary physician on the patient, but rather as the work of a team within a community. In order to follow these Cuban-trained Bolivian medical practitioners, I went to their places of work: clinics, ophthalmological centers, and hospitals. In addition to these groups of Bolivian health professionals, I conducted interviews with Cuban doctors who were training Bolivian stu-dents at Hospital Francisco and with directors of the medical diplomacy pro-gram at the Cuban embassy and the offices of the medical diplomacy program in Bolivia.

Global Health, Medical Education, and the Formation of Clinical Subjectivity

I locate this work at the intersection of literature in the anthropology of glob-al health, the anthropology of medical education, and recent work on the "clinical subjectivity" of the medical practitioner. The anthropology of global health focuses on instances when health becomes an aspect of international relations, with an "increased sensitivity to transnational social, ideological, political, environmental, and economic processes" (Janes and Corbett 2011: 136). Studies of "global health diplomacy" examine how ideas about health circulate, take root, and impact local environments, with work on changing global connections and transnational economic and political influences (Adams et al. 2008; Clarke et al. 2003; Petryna 2009; Redfield 2008). Examinations of medical humanitarian aid as an aspect of global health diplomacy have often focused on North-South humanitarian engagements (Bornstein 2003; Redfield 2005; Fassin and Vasquez 2005; Redfield 2008; Englund 2006; Ticktin 2006). Previous work has made clear that Cuba's medical diplomacy, which involves extended medical missions and medical education programs in over 60 countries around the world as a kind of South-South humanitarian engagement, follows a different logic than the aid engagements primarily discussed in the global health literature (Feinsilver 1993). Briggs and Mantini-Briggs (2010) and Muntaner et al. (2006) provide important examples of this work in their studies of the long-term and expansive Cuban-run *Barrio Adentro* project, where clinics run by Cuban doctors have been established in communities across Venezuela.

An aspect of these studies in global health involves attention to the circulation of health experts and expertise through humanitarian aid projects (Redfield 2005, 2008; Fassin 2007). However, in examinations of the flow of medical expertise as part of this diplomatic medical engagement, studies of Cuban medical diplomacy have often focused on the circulation and expertise of Cuban doctors who provide aid through medical missions to other countries (Briggs and Mantini-Briggs 2010; Feinsilver 1993, 2006). In this chapter I shift the focus to the foreign medical students and health practitioners who were trained by Cubans and returned to their home country. I track their knowledge production through this diplomatic engagement. An analysis of the discourses and practices these students developed and applied within the context of their medical education and medical practice provides an essential perspective in a study of the global circulation of ideas about health and models of health.

I extend this examination of the formation and circulation of experts through a global health diplomacy program with a focus on the anthropology of medical education and clinical subjectivity. In the literature, a number of scholars have examined the socialization of medical students and health practitioners into the field of medicine (Kleinman 1988; DelVecchio Good and Good 1989; Good 1993; Good and DelVecchio Good 1994; Davenport 2000). I build on this work by examining the way the socialization of these medical students was framed as a transformation process that I argue is an essential part of the development of clinical subjectivity. However, in the Cuban-Bolivian case discussed, this transformation was not just a process of socialization from layperson to medical practitioner but was also entangled in discourses of nationalism and border crossing. For example, Bolivian students trained in Cuba explained how they learned and made mobile new ideas about health, medicine, community, and medical practice, and expressed the impacts of working transnationally, both cultural and medical, on their experience of becoming health practitioners.

Other work has examined the psychology of the medical expert (Fox 1957), the phenomenological conditions of the medical student's life world, and the impact of international medical training on medical students (Wendland 2010). Byron Good (1994), for example, asserts, "Learning medicine is not simply the incorporation of new cognitive knowledge, or even learning new approaches to problem-solving and new skills...Learning medicine is developing knowledge of this distinctive life world and requires an entry into a distinctive reality system" (Good 1994: 71). Good argues that by extending these same categories to an analysis of the medical expert-in-training we can understand "how the medical world gets built up as a distinctive form of reality for those who are learning to be physicians" (Good 1994: 67). Following Stonington et al. (2011), I examine the processes of medical education within the context of the growing focus on global health (Holmes et al. 2011: 106), since an aspect of Cuba's diplomatic relations with Bolivia is the "production of clinical subjectivities" (Holmes et al. 2011: 106).

To do this, I take Mary-Jo DelVecchio Good's (1994, 2011) idea of the "clinical narrative" that doctors create about patients and follow Stonington (2011) by turning Foucault's outward medical gaze "inward" to look at the clinical narratives Cuban-trained Bolivian medical students and health practitioners construct about themselves and their processes of becoming health professionals within a transnational diplomatic engagement. I follow Holmes et al. (2011), who question, "What kinds of people are formed through contemporary processes of clinical training, and how do these evolving subjects transform health, power, and other aspects of social life?" (Holmes et al. 2011: 2). They suggest that medical education involves processes of self-formation and self-crafting on the part of the medical student or physician, creating a counterpoint to Foucault's emphasis on the medical gaze that focuses on and creates the patient as an object of medical attention (Holmes et al. 2011: 4; Foucault 1994). As such, I examine the personal and practical manifestations of Bolivian medical practitioners' unique experiences in medical education in Cuba in order to question the more subtle ways that ideas about health and health care transform within the context of transnational movement, as well as in parallel with larger national political processes.

Background: Cuban Medical Diplomacy in Bolivia and the Bolivian Health Context

Over the last 50 years, the Cuban state has engaged in a unique form of medical diplomacy, or the provision of medical aid and education in exchange for both symbolic and material capital to over 60 nations around the world (Feinsilver 1993). To reinforce this diplomatic work, in 2004, Venezuelan president Hugo Chavez initiated the ALBA agreement (*Alternativa Bolivariana para las Américas* or the Boliviarian Alternative for the Americas) between Venezuela and Cuba as a stated alternative to the U.S. "Free Trade Area of the Americas" (FTAA) (Arreaza 2004). ALBA emphasizes alleviating the social and economic disparities between member nations with the aim of eradicating poverty. ALBA has since been joined by Bolivia, Nicaragua, Ecuador and many smaller Caribbean nations and has expanded to include projects in other areas of social and economic need such as agriculture, military aid, and educational expertise. Bolivia joined ALBA in 2006 after Evo Morales, Latin America's first indigenous president, was elected (Smith and Obiko 2006).

Cuban doctors began working in Bolivia in 2006 during severe flooding in the Beni Department. The first medical brigade arrived with 140 doctors who had been trained to give humanitarian aid during disasters and catastrophes. The Bolivian government asked the Cuban doctors to remain in the country, and since then Cubans have mainly provided primary medical attention. Most notably, the Cuban medical presence in Bolivia includes *Misión Operación Milagro*, an ophthalmological program with six centers around the country that perform ten different types of surgeries for Bolivians free of charge, as well as *Misión Moto Méndez* that helps to rehabilitate people with physical disabilities (Interview 2011). The newest aspect of Cuban aid in Bolivia involves the development of human resources through the provision of scholarships to Bolivian students to study in Cuba. At the time of this research, there were approximately 5,000 students being educated and trained between Cuba and Bolivia. Of these, there were around 1,600 students who started their studies in Cuba and were finishing their medical education in Bolivia under the instruction of the Cuban medical brigade in Bolivian institutions, such as Hospital Francisco (*Ministerio de Salud y Deportes* 2012).

In a recent Bulletin from the Bolivian Ministry of Health about the six-year anniversary of Cuban-Bolivian cooperation, the Minister of Health and Sports, Dr. Juan Carlos Calvimontes, noted, "The presence of Cuban doctors changed the life of our country. They have not only guided and helped us, they are also giving a new concept of public health in Bolivia" (*Ministerio de Salud y Deportes* 2012). However, the question of Cuba's influence in Bolivia and what exactly "travels" through medical diplomacy is contested. For example, one Cuban doctor at Hospital Francisco told me, "In the hospital here we attend to patients and teach medical students how to practice medicine, but we do not teach a mode of thinking." This doctor emphasized that political ideologies about health were not part of the medical curriculum for Bolivians trained by Cubans, and highlighted the inherent contextual differences between the nations. Yet, Dr. Gonzalez, a Cuban psychiatrist and the subdirector at Hospital Francisco, provided a different perspective. He explained that they were incorporating aspects of the Cuban model into their work at Hospital Francisco and were applying it, for example through prevention and promotion programs. He felt the Bolivian people would benefit from aspects of the Cuban model, especially in the ways doctors interact with patients in the clinic, which he felt could also lead to a "change in mentality amongst patients." To do this, Dr. Gonzalez explained, doctors at Hospital Francisco engaged in outreach programs, such as health fairs and home visits, to teach people about this community mentality. He told me, "We take into account the environment of a community and the idea that man is not a solitary being, he is part of a family and part of a society."

In the following sections, I narrow the focus in this diplomatic relation to the Cuban-trained Bolivian medical students and health practitioners to examine how these ideas about health, care, and community developed through their studies in Cuba. I question how discourses and practices related to health care circulated between nations and became part of a diplomatic engagement.

Transformation Narratives

During my research in Cuba and Bolivia, I found commonalities in the ways Cuban-trained Bolivian medical students and health practitioners narrated the personal transformations they experienced through their training in Cuba and their return to Bolivia. They emerged with what, drawing on Stonington (2011), I consider to be a new sense of "clinical subjectivity," based around the approaches to health care provision that these health practitioners felt they developed through their studies in Cuba. By "clinical subjectivity," Stonington (2011) refers to a shift he saw amongst clinicians in Thailand when they began to reshape themselves inwardly based on their outward clinical gaze on patients during end-of-life care. Similarly, in my work, I understand clinical subjectivity as Bolivian health practitioners' inner personal and professional development based on a gaze outward through both their Cuban medical education and through their interaction with patients in the contexts of Cuba and Bolivia. Indeed, in interviews, Bolivians often connected their personal transformations with their transition across what they experienced as the national and cultural borders from Bolivia to Cuba and back to Bolivia. They specifically emphasized the transformations that occurred in how they understood medical training, health, care, the patient, and the idea of community. As such, learning and applying what they considered to be a "Cuban" approach to medical practice was often framed as a conversion experience, incorporating what I call "transformation narratives" of the health practitioner, or Cuban-trained Bolivians' narratives of their own conversion experiences through their medical training in Cuba.

Mary-Jo DelVecchio Good's (1989) idea of the "clinical narrative" examines the narratives that doctors create about and with their patients. Through my analysis of the development of clinical subjectivity, I look at a different kind of clinical narrative that involves the narrative a health practitioner creates about her own clinical subjectivity in relation to her experiences learning medicine. The idea of the "clinical narrative" is useful because many times the narratives of transformation that I heard emerged from clinical experiences of interaction with a doctor-teacher, a patient, or an aspect of the health care system in Cuba, and became part of the stories of personal and clinical development that these students and health practitioners told me. Often these experiences were presented to me with nearly religious fervor, where the Cuban-trained Bolivian health practitioners systematically glorified the Cuban health care system. Many of the transformations the Bolivian health practitioners experienced involved discussions of the differences they experienced between the Cuban and Bolivian health care systems in regards to doctor-patient interaction and conceptions of holistic, community medicine. However, treating patients in a holistic way is not just particular to the Cuban health system, though it is an approach that was not yet actualized in all areas of the Bolivian context. Students found it striking that this holistic approach to health was present in a systematic way in the Cuban health education programs.

When I met with Dra. Martinez, the director of Cuban medical education in Latin America, at the offices of the Cuban medical diplomacy program in Bolivia, she explained:

One of the things that foreign students talk about is how they felt they were changed or transformed, because many of them were very selfish when they arrived. It is not their fault because in their countries life is like this, you have to fight for what you want, and only watch out for yourself and your loved ones. But later they see this communal or collaborative work and they learn to worry about others and eventually they develop a solidarity sense of being: a sense of fellowship and an ethic of helping the one who is in trouble.

Dra. Martinez noted a personal transformation she witnessed in students who transitioned from "selfish-minded" to "communal-minded," which she associated with their education in Cuba. She emphasized the importance placed on the community orientation to medical practice that is part of medical training in Cuba, where the patients' social and community environments are considered part of their health indicators. This community sentiment was echoed in Ernesto's discussion of his experience in Cuba with which I opened this chapter.

Dra. Martinez also made another move to position a student's subjectivity prior to medical training in Cuba as a symptom of the tendencies of the conditions of that student's home country, emphasizing the impact she found the transnational movement to have on these students. She suggested that medical education in Cuba was not just about learning the technical skills of a doctor or health practitioner, but also held the possibility for a student's subjectivity to be reshaped through the immersion in Cuba's health care system. Adams and Kaufman (2011) similarly suggest this transformation when they note, "Re-skilling technologies and educational strategies mandate new ways of knowing patients, systems of service delivery and above all, the new kinds of ethical opportunities that clinicians need to embrace" noting the "entrenched moral economies of health care settings" (Adams and Kaufman 2011: 317). These transformation experiences marked the shift to a new kind of clinical subjectivity for these students that in many cases reoriented their understanding of what constitutes the patient, health, and healing. More broadly, it shifted their sense of the ethics of providing health care, as well as, crucially, a sense of community and national responsibility. In their narratives they referred to some of the educational strategies that were essential in this shift, including teacher-student interaction, doctor-patient interaction, and a shift from individualism to a sense of community.

The approach to transforming medical education and clinical subjectivity was essential to the development of the Cuban health care system. Following the 1959 revolution in Cuba, the government asserted health as a right of the people and began to focus on preventive medical care within a socialist health service model. The transformation in health care in Cuba involved a shift in the role of the doctor, who going forward was expected to play a community leadership role, which included assessing the community and family environment and local perspectives on health, as well as living in the neighborhood of his or her patients. Living in the same neighborhood as their patients allowed doctors and nurses the ability to "immerse themselves in the environment and in the psycho-social-biological problems of their patients and to provide immediate and continuous care" (Feinsilver 1993: 41). Cuban-trained Bolivians whom I spoke with emphasized the impact of learning this approach to practicing medicine.

For example, Teresa, a young Bolivian physiotherapist, spent five years studying in Cuba. When I asked her to describe her education learning in Cuba she emphasized what she called interchangeably the Cuban "sense of health" (*sentido de salud*) and "logic of health" (*lógico de salud*) (Gibbon 2009). For Teresa, there were three different experiences of learning this "logic" that stood out in her training, marked by her interactions with teachers and patients. She told me that the beginning stages of her training caused her quite a bit of irritation when she still did not fully understand what she called the "Cuban approach" to health care and doctor-patient interaction. She explained:

> I can tell you that I had a test in my third year, a doctor asked, "What is the name of your patient" and if you can't answer the question you get a zero…I thought to myself, "He is crazy, how is he going to fail me because of that?" I said, "Her name is Marta." And after the test I asked myself why did he ask me this? Later, I went to the doctor's office and I told him, "It's not fair that you were going to fail me because of that." He said, "You are not treating objects and you are not treating illnesses. You have to treat patients using their names." That was very important for me. So, I never say, "The patient who is in bed number 5 has this or has that"; I always say, "Juanita, who is 32 years old, has this."

Teresa's second moment of recognition occurred a bit later in her training. She recounted:

> I had these teachers who would show me an x-ray and they asked, "What do you see here?", and I said, "I can see this, I can see that." And then they would ask, "What is the best treatment?", and I used to say, "Well they need this or they need that," and then my teacher said, "How can you dare to give

a treatment if you have not seen the person." I used to think, "oh god again, that."

Finally, Teresa told me about what she considered to be the key moment of her transformation.

A very deep experience I had was when the doctor told me I was going to have my first patient and I did the interview very fast, and I thought, "Oh I did it very well." Then I saw my second patient and I did it even faster, and I felt "Oh, I'm excellent." But the teacher told me, "Now you have to sit and listen, and listen to everything that the patient has to say because you have all the time in the world to do that" even though I thought, "But I can do it faster." So I did what the doctor said with the next patient, and the woman started talking about this and that, and her son, and that she was not feeling well. And I thought, "When is she going to tell me what is the problem, what hurts?" But she kept talking. Finally the patient said, "I felt so good with you today that nothing hurts anymore." In that moment, the teacher looked at me and said, "The only thing that I ask you to do is to listen."

For Teresa, this series of experiences first seemed irritating until she reached a transformative moment where she felt this logic of medical care was able to reshape her conception of health care provision and what constituted the relationship between doctor and patient. I asked Teresa how this transformation impacted her work in Bolivia and she told me that it has mostly shaped her ideas about how to relate to patients. She explained:

I try not to have any hierarchy between myself and the patient. I don't want my patients to look at me as if I was owning their bodies, because what I saw in Cuba was that the patient is at the same level as the doctor. The doctor would talk to the patient in "*tú*" informally. That is something we don't have here in Bolivia, here it is like the power of knowledge, and that is the way doctors are regarded as if they have power over knowledge. I want my patients to call me Teresa, not *Doctora* or *Licenciada*.

Another health practitioner, Verónica, also associated the development of her clinical subjectivity with an interaction with a patient. Verónica received a fellowship from the Cuban government to spend five years studying health technology and public health sciences in Cuba. She told me that one day, after observing Cuban doctors interact with patients, she realized she could not be cold or it would hurt the patient. She told me that she talked about baseball with her patient to help the patient gain trust in her. After the exam, the patient invited her to her house, gave her clothing, and they became friends. Verónica focused on how this experience shaped her current work:

> In my daily practice I treat the patient with respect but also as if she was my friend. I understand she might be afraid of what I am going to say or do, so I try to have more interaction with her and check in with her about how it is going. I don't want her to come to the consultation afraid or worried that she is going to be in pain, so…I try to talk to her and make it fun for her.

This sentiment was repeated in nearly all of my interviews, as health practitioners explained the importance of laughing or joking with their patients or making the visit comfortable and fun. As one medical student at Hospital Francisco expressed, "A smile can improve the patient." In their transformation narratives, Teresa and Verónica focused on particular practice-based clinical experiences that involved dynamic relationships with patients.

Cuban-trained Bolivians also noted a change in their personal world-views through their experience of learning in Cuba, which impacted the way they saw themselves as health practitioners. For example, Mirna, a Bolivian epidemiologist and medical technician, explained that prior to her studies in Cuba she held an "individualist" view of the world, where she placed herself and her family at the center, what she said was very "closed-minded." However, while in Cuba, Mirna explained that she started thinking more collectively and that the types of relationships she had with people upon her return – patients, colleagues, friends, and family – were "much more humanitarian." She explained, "Even though Bolivia is plurinational, there is a language of love between people that is much stronger or more than just thinking about distinct groups or languages." Mirna's transformation experience in her clinical subjectivity in Cuba was paralleled by a religious conversion experience. She explained that her family is Catholic but it was only in Cuba that she had her first communion. In Bolivia she had chosen not to receive communion because when she went to the church to get communion, the first question they asked her was whether she went to a private or public school, and she felt, "'Why should there be a difference?' In Cuba," she told me, "the church didn't ask me this kind of question and just said 'tell us when you are ready.' The Catholic Church in Cuba is another form of community and there are no differences between whether you went to a private or a public school, because all participate in the religious acts." Mirna suggested, "This vision of the church in Cuba, based on community and participation, is like the Cuban medical brigades with doctors and other professionals all working together." Mirna became emotional as she described this double conversion and took out a small picture of the Bolivian image of the Virgin of Copacabana; she told me that she carried the image with her to remind her of her experience in Cuba. Unlike Teresa and Veronica, whose transformation experiences involved direct interactions with a teacher or a patient, Mirna's experience involved a transformation in the way she understood herself in relationship to others, shifting from a discourse of individualism to one of community and participation.

The shift in subjectivity that these health practitioners noted involved a shift to being a specific kind of doctor *because* they were trained in Cuba. They emphasized their experiences through discussions of the medical interaction between physician and patient, with a focus on relationship building as equally if not more valuable than technical or specialty expertise. They also noted a shift in their thinking about the patient as an individual to seeing the patient as part of a community, bringing a public health orientation into their medical practice and the medical outreach projects that they created and engaged in.

Discursive Contrasts: Negotiating Distinct Sites of Medical Practice

The transformation narratives discussed above and the resulting sense of clinical subjectivity that these medical students and medical practitioners felt they developed marked a transition to a medical practitioner with a different way of envisioning health provision. This vision of health provision encompassed their relationships to patients and their positionality, or subjectivity, in a clinical and community context. Stonington notes a similar shift in clinical subjectivity through the historical introduction of palliative care to Thailand which, in confluence with Buddhism, led to transformed practices of medicine, particularly in the relationship between physicians and patients (Stonington 2011). However, because of the transnational context of Cuban-trained Bolivians' education, I found that their transformations were further emphasized in the discursive contrasts Cuban-trained Bolivians drew between what became framed as distinct sites of medical practice. This framing allowed them to feel that they could package something from their training in Cuba to bring back to Bolivia. Through these contrasts, the Bolivian health practitioners positioned themselves as those who could do boundary work between what was set up as two bounded systems by attempting to bridge the approaches to health care provision and the social and political contexts that they felt each system encompassed (Gieyren 1983). While these practitioners produced very stark contrasts and at times came across as the harshest critics of the Bolivian health care system, they also placed themselves at the boundaries of these "worlds" and negotiated these boundaries through their medical practice. I found that the primary contrasts used by these Bolivians revolved around health care system structure, styles of medical training, doctor-patient relations, and cultural context, mirroring some of the areas that played key roles in their personal transformations discussed above.

In a transnational medical education context, the health practitioners I spoke with noted that they had to figure out how to transgress the "two worlds" through their transition from Bolivia to Cuba and back to Bolivia again. Cuban-trained Bolivians bounded their experiences of "Cuba" and "Bolivia" and created contrasts to help give their experience meaning back in Bolivia. They illuminated aspects of their transformation through juxtapositions with differences or limitations they found practicing medicine in Bolivia. Through these contrasts, Bolivia was often framed by these students and medical practitioners as "underdeveloped" or "pre-modern" in the economy of knowledge surrounding medical practice (Pigg 2001). Similarly, Brada (2011) shows in a medical education setting in Botswana how contrasts were used to create a "spatio-temporal" distance between practicing medicine in Botswana, deemed to be "like going back in time," compared to the framing of the United States as "'having moved away from' such practices" (Brada 2011: 296). While these contrasts positioned Cuba ahead of Bolivia in the "spatio-temporal" framework of health care system development, the Cuban-trained Bolivians also use the contrasts to root their work in the Bolivian context and instigate their work for changes within Bolivia.

Many of the contrasts that Cuban-trained Bolivians made during our discussions focused on broad issues about the structure and philosophy of the current health care systems in Cuba and Bolivia. For instance, while in Cuba, Ernesto and Johana suggested that at a system-wide level, the health care system in Bolivia had not undergone the kind of drastic revolutionary change that Cuba went through. They felt that without these large-scale changes in health system ideals in Bolivia it would be more difficult to implement the approach to medical care they learned in Cuba. The students and health practitioners also created contrasts between the approach to clinical training in Cuba and the approach that they found to be predominant in Bolivia. Alejandra, a medical student at Hospital Francisco, for example, explained her surprise at her Cuban teachers' approach to pedagogy in school, "They were very open and the teachers don't guard information or knowledge. Here [in Bolivia], I've seen doctors who hold back information and are very egotistical." Other medical students explained that they began attending to patients within their first year of medical school, while in the Bolivian medical school system students only enter practical training in the third or fourth year of medical school. Pedro commented, "Because of this, Bolivian doctors enter the clinics as children." Gabriela added, "In Cuba we learn to be general doctors first, and later can add a specialty, while in Bolivia there are many more specialists because it brings in more money."

Further, just as doctor-patient interaction was an important site of transformation for many of these health practitioners, it also became a topic around which some of the greatest contrasts were drawn. Alejandra compared Bolivian doctors to Cuban doctors:

> Doctors here don't explain treatments to patients, while Cuban doctors explain treatments and medications and teach with psychological support. An important part of my job is to explain health issues and medication procedures to patients in a detailed way so that they will understand what their illness is and what they need to do. I can't just give pills and walk away since, for example, with diabetes, it requires not just pharmacological treatment but also diet and social treatment.

Further, she explained, "I think that the Bolivian people don't accept treatment as easily as Cubans or are not accustomed to seeing a doctor. So, when the doctor explains everything and how it impacts their lives it really helps with the treatment."

One of the most common contrasts made by the Cuban-trained Bolivians involved the practice of house visits that is common in Cuba. Medical student Daniela explained that in Cuba, people were always comfortable opening their doors to the doctors. She said, "Now in Bolivia, we work at a small hospital but would like to expand and increase our work. When we went door to door in Cuba the families always opened their doors for us, but here they are more closed and it is harder." Juan Carlos, another medical student, explained that in the Cuban program they start out in clinics and always visit the community. However, he explained, "In Bolivia when you knock on the door and ask about their sicknesses they don't open the door because they don't have faith in the doctors." However, the Bolivian medical students noted that this reluctance is related to cultural context between the two countries, and raised the issue of the attention to cultural differences that is required in Bolivia based on the diversity of the population. Pedro explained, "Bolivia is a multiethnic, multicultural country. Each culture has its own models, ethics, and ideas about health. It is very difficult to understand culture and to treat health, since it can't just be about imposing our own ideas about health onto people here."

Drawing these contrasts between Bolivia and Cuba became a way that Cuban-trained Bolivian medical students and health practitioners demonstrated some of the challenges that they encountered when they returned to Bolivia and tried to implement their training in what they felt to be a very different cultural and political context. However, they also used these contrasts as a way to mobilize the boundary work they engaged in, discussed below, in order to create bridges between the system they were trained in and the system in which they practiced.

The "Social Work" of Bringing Cuba to Bolivia

The question of national and personal development intertwined for these health practitioners, as their medical education in Cuba instilled many with a tremendous sense of both personal agency and national responsibility. However, while the Cuban-trained Bolivian medical students and medical personnel I spoke with emphasized their aspirations to participate in Bolivia's health environment, they were not always able to officially do so. Some complained that their work was not officially recognized in Bolivia. Many told me that they had tried to meet with officials at the Ministry of Health to talk about helping with the health efforts, but had not gotten any response about what they could do. Others also had trouble getting their titles earned in Cuba recognized in Bolivia. Teresa expressed this frustration and her own solution, "Yeah, of course when you are flying back to Bolivia you are full of hope. But then you can see that here there are so many doors that are closed by the government. Then you are disappointed, thinking, *what do I do now?* At least you can still give your patient quality treatment."

Amidst difficulties encountered, many found informal ways of bringing their Cuban-learned approach to health to Bolivia through social projects and aspects of their medical practice. Many of the medical students and medical practitioners I spoke with expressed the desire to manifest their new understandings about health and health care provision through engaging in social outreach projects or community projects, what they called "social work" (*trabajo social*) as a way to bring Cuba to Bolivia. For example, Teresa told me, "I had the chance to stay in Cuba, it wasn't a very serious chance, but I thought, *'no, I'm going back to my country because I have a kind of a social and moral debt to Bolivia. I am going to treat my people in Bolivia as I learned to treat people in Cuba.'"* Some noted the ways they brought their Cuban training directly into their interactions with patients and other colleagues. Others found ways to begin to bridge the differences between their Cuban training and the experiences they encountered in Bolivia, with a focus on the particularities of Bolivia. Alejandra explained:

> One of the reasons we are here now to complete the studies we began in Cuba is to learn about the style of medicine in Bolivia and to implement it in a practical way along with what we learned in Cuba. We are also here to learn about culture and implement an approach to culture in our practice. In this country it takes a lot of time for change, but the family doctor has an important role in helping to create these changes for the Bolivian society.

Health practitioners I spoke with also created projects to work with the Bolivian community in ways that reflected aspects of approaches that they developed in Cuba. For example, epidemiologist Mirna worked for the Ministry of Education and was involved in training Bolivians in a rural town three hours from La Paz in epidemiology and health sciences. She woke up at 5:30 every morning to make the long journey on public transportation and was creating her own textbooks and materials based on the materials she received in Cuba. José, a Cuban-trained Bolivian ophthalmologist, was in the process of creating a mobile eye clinic that would be able to go to rural communities that do not have clinics. He explained that his experience in Cuba helped him to come up with this idea. José also told me that he gave free care to people with difficult economic situations. Others had ways of bringing their training in Cuba into their personal practices. For example, Verónica taught her colleagues about her approach to patients, though she noted, "Sometimes they don't understand and even patients are sometimes uncomfortable with the informality."

These projects and approaches to medical care were some of the ways that the Cuban-trained Bolivians packaged what they considered to be aspects of their "Cuban" training and made mobile ideas about health and health care provision to implement in Bolivia. They used these projects to bridge the contrasts that they created between Cuba and Bolivia. These practices occurred not only at the level of doctor-patient interaction, but also through social projects that mirrored some of the changes that the Bolivian government was implementing as part of the health care reform. Further, this "social work" also became a way for students and health practitioners to negotiate the national and cultural boundaries that they drew between Cuba and Bolivia by engaging in projects and approaches to health care provision that integrated aspects of what they learned in Cuba while keeping in focus and grounded within the context of the Bolivian population and health care system. For the Cuban-trained Bolivians I spoke with, their transformation to medical practitioners through a transnational medical education program instilled them with a tremendous sense of both personal agency and national responsibility, which sometimes was obstructed when they returned to Bolivia. In this way, questions of nation and personal and professional development continually intertwined in the Cuban-Bolivian medical diplomacy engagement, informing and shaping the new clinical subjectivities of these medical students and health practitioners.

Conclusion

I have shown how medical education abroad was understood by Cuban-trained Bolivian health practitioners as the entrance into a different medical culture. They found this experience to be personally transformative because of the powerful way it was able to reform their conceptions of medical practice and care. It also offered them a glimpse of different possibilities in their practice within their own national context. The transnational formation of health practitioners manifested as both a personal transformation and as a framework for health practitioners to position themselves within larger-scale changes in the Bolivian health care system. I thus found that questions of nation and personal and professional development were inextricably related in the Cuban-Bolivian medical diplomacy engagement.

These different narratives of transformation – personal, professional, and national – and the resulting practices that emerged show not just a straightforward story of the politics of Cuban medical diplomatic relations or the transfer of health ideologies across national borders. Rather, they allow for a narrowed focus on the ways that Bolivian health practitioners' clinical subjectivities were profoundly changed in the process of transnational medical education and how these changes influenced the ways they approached their practice and politics within Bolivia. I trace the various ways in which personal transformation and ideas about health care emerged and were bounded in a nationalized trajectory or set of goals for Bolivian healthcare, such as approaching patients in a familiar manner and participating in community health outreach projects rather than focusing only on the individual patient in a hospital setting. I found that this educational system did not just provide a wholesale import of medical infrastructure and expertise from Cuba; rather, it allowed for a reconfiguring of the terms in which ideas and practices relating to health and health care systems were understood and practiced. This analysis provides a way to examine some of the intended and unintended outcomes of the Cuban medical diplomacy program when aspects of its ideological premises are embodied in and reframed by the mobile health practitioner.

References

Adams, V., Novotny, T., and Leslie, H. (2008) 'Global health diplomacy', Medical Anthropology, 27.4: 315-323.

Adams, V. and Kaufman, S. (2011) 'Ethnography and the making of modern health professionals', Culture Medicine Psychiatry, 35: 313-320.

Albro, R. (2005) 'The indigenous in the plural in Bolivian oppositional politics', Bulletin of Latin American Research, 24.4: 433-453.

Arreaza, T. (2004) ALBA: Bolivarian Alternative for Latin America and the Caribbean. Online. Available HTTP: <venezuelanalysis.com/analysis/339> (accessed 21 June 2011).

Bornstein, E. (2003) The Spirit of Development: Protestant NGOs, Morality and Economics in Zimbabwe, New York: Routledge.

Brada, B. (2011) '"Not here": making the spaces and subjects of "Global Health" in Botswana', Culture Medicine Psychiatry, 35: 285-312.

Briggs, C. and Mantini-Briggs, C. (2009) 'Confronting health disparities: Latin American social medicine in Venezuela', American Journal of Public Health, 99.3: 549-555.

Brotherton, P.S. (2005) 'Macroeconomic change and the biopolitics of health', Journal of American Anthropology, 10.2: 339-369.

Clarke, A., Shim, J., Mamo, L., Fosket, J. and Fishman, J. (2003) 'Biomedicalization: technoscientific transformations of health, illness, and U.S. biomedicine', American Sociological Review, 68.2: 161-194.

Danielson, R. (1977) 'Cuban health care in process: models and morality in the early Revolution', in S. Ingman and A. Thomas (eds.) *Topias and Utopias in Health*, The Hague: Mouton Publishers.

Davenport, B.A. (2000) 'Witnessing and the medical gaze: how medical students learn to see at a free clinic for the homeless', *Medical Anthropology Quarterly*, 14.3: 310-327.

DelVecchio Good, M.J. (1994) 'Oncology and narrative time', *Social Science and Medicine*, 38.6: 855-862.

—— (2011) 'The inner life of medicine: a commentary on anthropologies of clinical training in the twenty-first century', *Spring Science+Business Media*.

DelVecchio Good, M.J. and Good, B. (1989) 'Disabling practitioners: hazards of learning to be a doctor in American medical education', *American Journal of Orthopsychiatry*, 59.2: 303-312.

Eckstein, S. (1994) *Back from the Future: Cuba Under Castro*, Princeton: Princeton University Press.

Englund, H. (2006) *Prisoners of Freedom: Human Rights and the African Poor*, Berkeley: University of California Press.

Fassin, D. (2007) 'Humanitarianism as a politics of life', *Public Culture*, 19.3: 499-520.

Fassin, D. and Vasquez, P. (2005) 'Humanitarian exception as the rule: the political theology of the 1999 tragedia in Venezuela', *American Ethnologist*, 32.3: 389-405.

Feinsilver, J. (1993) *Healing the Masses: Cuban Health Politics at Home and Abroad*, Berkeley: University of California Press.

—— (2006) 'Cuban medical diplomacy: when the left has got it right', *Foreign Affairs en Español* under 'La diplomacia medica cubana: cuando la izquierda lo ha hecho bien', 6.4: 81-94.

Foucault, M. (1994) *The Birth of the Clinic*, New York: Vintage Books.

Fox, R. (1957) 'Training for uncertainty', in R. Merton, G. Reader, and P. Kendall (eds.) *The Student Physician*, Cambridge: Harvard University Press.

Gibbon, S. (2009) 'Genomics as public health? Community genetics and the challenge of personalized medicine in Cuba', *Anthropology & Medicine*, 16.2: 131-146.

Gieryn, T.F. (1983) 'Boundary-work and the demarcation of science from non-science: strains and interests in professional ideologies of scientists', *American Sociological Review*, 48: 781-795.

Good, B. (1994) *Medicine, Rationality, and Experience: An Anthropological Perspective*, Cambridge: Cambridge University Press.

Good, B. and DelVecchio Good, M.J. (1993) 'Learning medicine: the

construction of medical knowledge at Harvard Medical School' in
S. Lindenbaum and M. Lock (eds.) *Knowledge, Power, and Practice: The
Anthropology of Medicine and Everyday Life,* Berkeley: University of
California Press.

Gustafson, B. (2009) *New Languages of the State: Indigenous Resurgence and the
Politics of Knowledge in Bolivia,* Durham NC: Duke University Press.

Heredia, N. (2006) 'Equidad y determinantes sociales de la salud' paper
presented at the 59th Asamblea de la Organización Mundial de la Salud,
Geneva, May 2006. Online. Available HTTP: <http://www.sns.gob.bo/
index.php?ID=Inicio&resp=224> (accessed 30 June 2011).

Holmes, S., Jenks, A., and Stonington, S. (2011) 'Clinical subjectivation:
anthropologies of contemporary biomedical training', *Culture Medicine
Psychiatry,* 35: 104-112.

Janes, C. and Corbett, K. (2011) 'Global health', in M. Singer and P. Erickson
(eds.) *A Companion to Medical Anthropology,* 1st edn, New Jersey: Wiley-
Blackwell Publishing, Ltd.

Kleinman, A. (1988) *The Illness Narratives: Suffering, Healing, and the Human
Condition,* New York: Basic Books.

Ministerio de Salud y Deportes (2012, March 5) 'Brigada medica cubana
cumple sexto año de ayuda al pueblo boliviano', Año 3, Boletín No. 512.

Muntaner, C., Guerra Salazar, R.M., Benach, J. and Armada, F. (2006)
'Venezuela's Barrio Adentro: an alternative to neoliberalism in health
care', *International Journal of Health Services,* 36: 803-811.

Petryna, A., Lakoff, A. and Kleinman, A. (2009) *When Experiments Travel,*
Princeton: Princeton University Press.

Pigg, S. (2001) 'Language of xex and AIDS in Nepal: notes on the social
production of commensurability', *Cultural Anthropology,* 16.4: 481-541.

Redfield, P. (2005) 'Doctors, borders, and life in crisis', *Cultural Anthropology,*
20.3: 328-361.

— (2008) 'Sacrifice, triage and global humanitarianism', in T. Weiss and M.
Barnett (eds.), *Humanitarianism is Question: Politics, Power, Ethics,* Ithaca:
Cornell University Press.

Smith, F. and Obiko, S. (2006, April 30) 'Cuba, Bolivia, Venezuela reject U.S.
trade', *New York Times.* Online. Available HTTP: <http://www.nytimes.
com/2006/04/30/world/americas/30iht-web.0430trade.html>
(accessed 28 June 2011).

Stonington, S. (2011) 'Facing death, gazing inward: end-of-life and the
transformation of clinical subjectivity in Thailand', *Culture Medicine*

Psychiatry, 35: 113-133.

Ticktin, M. (2006) 'Where ethics and politics meet: the violence of
    humanitarianism in France', *American Ethnologist*, 33.1: 33-49.

Waitzkin, H., Wald, K., Kee, R., Danielson, R. and Robinson, L. (1997)
    'Primary care in Cuba: low- and high-technology developments pertinent
    to family medicine', *Journal of Family Practice*, 45.3: 250-258.

Wendland, C. (2010) *A Heart for the Work: Journeys Through an African Medical
    School*, Chicago: University of Chicago Press.

Endnotes

1   Names have been changed to protect the privacy of research participants.
2   Name has been changed to protect the privacy of the hospital.

7

Cuban *Internacionalistas*, Sports, and the Health of the "Socialist Body" in Cuba and Africa

Ayesha Anne Nibbe

Cuba sends thousands of professionals overseas every year to serve as doctors, professors, and athletic coaches in far-flung places all over Latin America, the Caribbean, Asia, and Africa. These professionals, called *internacionalistas*, play a central role in bringing health to the "socialist body" – that is, they promote a sense of economic, political, social, and symbolic well-being within the socialist body politic at global, national, and household levels. While most discourse in the media focuses on the deployment of Cuban doctors as a form of foreign aid and diplomacy, the *internacionalista* presence extends far beyond hospitals and clinics. Sports professionals in particular play an important role in shaping Cuba's distinct brand of socialism and international engagement. While medical professionals engage in maintaining and restoring the "socialist body" to health by treating illness, sports professionals create and strengthen the health of the Cuban "socialist body" through particular forms of community- and state-building, through mentoring relationships that emerge in the intimate somatic regimes of coaching, and through buttressing the Cuban socialist state-economy by dispatching globally marketable sports professionals.

This paper illustrates the role *internacionalista* sports professionals play in promoting the health of the Cuban "socialist body" through an examination of ethnographic data collected both in Africa and Cuba. I studied a group of Cuban *internacionalistas* assigned to work in Uganda between 2006 and 2008. I later followed up that research with a subset of this group in Cuba after their return from Africa during the summer of 2009.[1] Cuban *internacionalistas* in Uganda hailed from various professions: doctors, nurses, mathematicians, food scientists, chemists, and musicians; however, I mainly observed athletic coaches and sports professionals in their daily lives, in their work setting at the university, and in training sessions on athletic fields in northern Uganda. I start by providing a theoretical framework for understanding the "socialist body," including a brief general historical background to Cuban *internacionalismo* in Africa (with a focus on Uganda). Ethnographic data illustrates how athletes play a key role in reifying, building-up, and maintaining the health of the "socialist body" in the Global South, the national "socialist body" in Cuba, and the "socialist body" located within the Cuban domestic household.

Internacionalismo and the "Socialist Body"

In his spoken autobiography, *My Life,* Fidel Castro recounts how Cuba first started sending *internacionalistas* to the African continent in the 1960s:

> despite the fact that imperialism had just lured half of our doctors away, leaving us just 3,000 [for the entire population], several dozen Cuban doctors were sent to Algeria to help the people there. And that was the way we started, 44 years ago, what is today an extraordinary medical collaboration with the nations of the Third World. ...we now have more than 70,000 doctors, plus another 25,000 young people studying medicine, and that does, without question, put us in a very special place – an incomparable place, and I am not exaggerating in the slightest – in the history of humanity (Ramonet and Castro 2010: 310).

Since those early years of sending doctors to Algeria, overseas dispatches of Cuban professionals expanded to other fields, including university professors, teacher training professionals, and "cultural" *internacionalistas* like musicians, sports professors and athletic coaches.[2] In some cases, sports professionals have been the dominant presence overseas; for example, Uganda hosted a larger contingent of Cuban athletic coaches and sports professors and fewer Cuban doctors.[3] One *internacionalista* in Uganda explained why sports are such an important component of *internacionalismo:* "Sportsmen improve the population, and encourage people to make exercise for health, to build strong physical capacity and prepare people for the life."

Cuba has promoted sports to foster political agreement within working classes since the beginning of the revolution. Cuban Revolutionary officials were "only too glad to support and enhance a sport [baseball] to which the masses were so passionately attached" (Arbena 1988: 131).[4] While the revolutionary ideal of widespread citizen access to sports facilities is largely unrealized, Cubans are still avid spectators of sports on television and in stadiums. Admission prices are set low enough to facilitate regular attendance at athletic competitions.[5] Sports are seen as an important way to develop the ideal Cuban national character – athletic activity promotes a healthy population that is not only productive, but also physically prepared to fight outside aggression if necessary (Arbena 1988: 127) – a decidedly masculine tone to revolutionary "*cubanidad*" (Carter 2008: xi).

The Olympic Games are particularly important to Cuban *internacionalismo*. The Games provide a world stage upon which Cuba can demonstrate that its socialist system not only works, but that it is measurably *better* than the Western capitalist system in the United States.[6] Prior to the revolution, Cuba was active in international sports competitions and was one of the early national participants of the modern Olympic Games. However, it was not until after Fidel Castro came to power in 1959 that Cuba raised its stature as a global Olympic power, eventually joining the ranks of the top ten medal-holding countries during the 1976 Montreal Olympic Games.[7] The Olympics are a critical site for Cuban ideological battle, not only against the United States, but also against global capitalist forces that increasingly permeate the Olympic Games.[8] Cuba invests considerable resources into improving athletic programs in the Global South by sending Cuban coaches to train their national teams in Olympic sports, particularly focusing on low-cost sports (for example, track and field, marathon running, boxing).[9] In countries like Uganda and Botswana where my informants work, Cubans were recruited specifically to bolster national standing in the Olympics.[10] The Olympic Games also provide a stage for Cuba to build one-on-one ties within the Global South; for example, during the Olympics in 2008, the Cuban coach of the Ugandan Olympic boxing team arranged sparring matches and social events with the Cuban team in the Olympic Village. By dispatching *internacionalistas* to raise the performance to international-competition levels in subaltern nations, Cuba maintains a symbolic foothold and continues to build strong ties in the Global South.

It is obvious that sending professionals overseas is a method by which the Cuban government builds symbolic capital and strengthens geopolitical ties within the Global South. However, the symbolism of sending doctors, sports professionals, and professors overseas is also a reflection of a distinctly Cuban socialist philosophy on developing the body in all senses – physical, mental, and social. This Cuban "socialist body" is produced not only through advancement of what is often called "human capital," the development of individual mental and intellectual capacity, particularly through education, but also by building up the individual physical form through the promotion of good health, physical strength, and endurance. In addition, this "socialist body" of local, national, and global community is strengthened through educational, sports, and medical institutions. The Revolution is built on the contention that the well-being of the individual bodies of the Cuban population is a "metaphor for the health of the body politic" (Feinsilver 1989, 2010: 66). And in a very blunt sense, the Cuban emphasis on sports is not only symbolic of the Cuban state as a healthy political body, but it is also a geographical embodiment of Fidel Castro himself. This Cuban "socialist body" lives out the principles of Castro's particular form of communism, and it is fully equipped and ready to embody – and physically fight for – those principles, ideals, and values in everyday physical form. Castro's physical form acts as the prototypical symbol of the Cuban nation, and his aging "becomes a metaphorical battleground for the staging of a visceral politics of the 'withering state'" (Brotherton 2008: 260).

The broad conceptual framework used in this chapter to discuss the "socialist body" is based on the Scheper-Hughes and Lock (1987) notion of a somatic form with three manifestations: the phenomenally experienced body; the social body through which nature, culture, and society are experienced; and the body politic. While all three forms are embodied in the individual form of the *internacionalista*, this paper focuses on the third aspect of the "socialist body" – that is, the *body politic* or the socialist body as a political space.[11] The health of the "socialist body" manifested as a political space takes on three different spatial configurations. The most expansive manifestation of this is the *global* "socialist body" located largely in the Global South, for which Africa plays a particularly important symbolic and strategic role.[12] Second, the health of the "socialist body" takes on a *national* form in the health of the Cuban nation-state. The last spatial configuration is the *local household* level of the body politic, the health of individuals collectively assembled into family groups under the collective of the Cuban household. At the household level of the body politic, the everyday, lived reality of the Cuban body politic is experienced, and therefore it is the site that reveals the challenges and contradictions of the "socialist body" at all three levels.

While all *internacionalistas* – doctors, professors, and sports professionals – bring health and vitality to the "socialist body" and provide a stage for ideological battle with the United States and global capitalism, sports professionals play a unique role in the reification of the "socialist body" through the micropolitics of Foucauldian bodily discipline. According to my informant, the goal of Cuban *internacionalismo* is "to transmit our culture [and] our knowledge," adding, "but for me [sports] is only an instrument. In the lessons or coaching sessions you transmit ideas about our political ideals." Through practiced and repeated movement, sporting bodies produce cultural meanings around citizenship and identity by creating an "inscriptive space" in which meanings about nationhood can be recorded (Wilson 1994: 266). In this way the State – which otherwise is a "mask" (Abrams 1988), a "fetish" (Taussig 1992) or an "imagined community" (Anderson 1991) – is physicalized and made legible through the imprinting of social values onto the somatic form through everyday bodily practice (Carter 2008: 3). Thus, the disciplinary aspects of athletics render the sporting body the site where the "body politic" is naturalized and reified. Moreover, the emotions elicited by athletic performance create "communities of memory" (Malkki 1997) that reinforce local, state, and global discourses and bring life to the body politic. Through sports, the "socialist body" – the Cuban domestic household, the Cuban State, and the Global South – becomes a living, breathing, physical "entity" composed of a citizenry that is able to contest opposing values on the local, national, or international stage. In Uganda, I witnessed Cuban sports professionals working very directly towards building Ugandan physical capacity: while their official job was coaching and teaching in the university, the Cuban *internacionalistas* voluntarily approached the local government in northern Uganda to launch a pilot project for primary school sports education. This pilot project caught the attention of a national political leader and was eventually written into a proposal sent to the Ministry of Education to revamp the larger public school system. Sports education, then, became one of the pillars of the educational apparatus in Uganda, and Cubans were able to export their somatic regimes directly into Ugandan society.

For playing this role, athletes get many privileges in the Cuban system. For example, university students who study in sports-related fields are given higher quality food, and they have access to luxuries like televisions while in the academy. Later on, they represent their country in national teams, are given permission to travel overseas, and later some athletes get tapped for special military duties that come with social prestige, like the serving in the Cuban Special Forces. (One informant told me that much of his national wrestling team contingent was sent to Angola in Special Forces units.) As a result, sports professionals, perhaps more than any of the other stripes of *internacionalistas*, tend to be more nationalistic and ideological about their roles of supporting and representing the Cuban revolution both within Cuba and overseas.

Internacionalistas and the Health of the "Socialist Body" in the Global South of Africa

Internacionalismo in Africa is a continuation of Ernesto "Che" Guevara's expedition to the Congo that aimed to unleash the former colonies from the chains of global oppression. Africa also plays a key role in the Cuban utopian vision of global socialist revolution, because Che and others in the revolution viewed it as the lynchpin of global inequality. If Africa could free itself from the unequal political and economic relations with the West, then the whole global capitalist system would theoretically unravel by breaking Western access to the rich African treasure trove of raw goods (e.g. cotton, fish), minerals (e.g. oil, diamonds, gold, coltan) and labor. The Cuban intervention in Congo was undoubtedly infused with the "ideal typification" that many Cubans hold about Africa and Africans, but Che was badly disappointed with the weak revolutionary fervor, lack of discipline, and pervasive "big man" syndrome that he found within the Congolese rebel leadership of the 1960s.[13] In fact, Che opened his diaries on the Congolese intervention with the words: "This is the history of a failure" (Guevara 2000: ix). While Che failed to realize his ideological and revolutionary goals in the Congo, Cuba has continued its relationship with Africa since then through aid, military intervention, and other assistance. Military assistance was the most prominent part of this relationship in Africa; along with Che's venture in the Congo, Cuba played critical roles in Angola and Guinea-Bissau in anti-colonial efforts against the Portuguese, in Algeria against the French, and in Mozambique.[14] Alongside these military expeditions into Africa, Cuba dispatched over 100,000 personnel, mostly doctors, to aid developing countries in Africa and Latin America.

In addition to these more prominent ventures, Cuba has strong ties across the African continent, including Uganda in East Africa. Uganda developed a solid relationship with Cuba in the context of Cuban *internacionalismo* after gaining independence from Britain in the 1960s. Just prior to sending the fledgling Ugandan national armed forces to the U.S. for training, Uganda deployed troops to Cuba for military exercises during the regime of the first Ugandan president, Milton Obote.[15] Cuba and Uganda established formal diplomatic relations in 1974 during the reign of Idi Amin, as part of Cuba's support for African nationalism and a bond fused by a shared aversion to their shared enemy, the United States.[16] At the beginning of President Museveni's triumph of his own guerrilla war and takeover of Uganda in the mid-1980s, Cuba was a key ally. However, by the late 1990s, with the fall of the Soviet Union and the commencement of the "Special Period" in Cuba, much of this Cuban military aid and assistance from Cuba to Uganda was reduced. Due to a decline in Soviet bloc funding in Africa, President Museveni strategically switched his political allegiances entirely to the West, in particular the United States.

Uganda is now seen as one of the darlings of the Western donor aid community, in large part because of the program of universal primary education[17] as well as President Museveni's quick and effective national response to the HIV/AIDS epidemic, policy initiatives that are aggressively pursued in Uganda up until today.[18] In the late 1980s and early 1990s, Uganda had an HIV national prevalence rate estimated to be 18-30% and rising ('Uganda AIDS' 2012).[19] Ironically, despite Uganda's dependence on Western (and particularly American) aid, it was Cuba, not the United States or the Western bloc that played an important role in calling attention to this escalating crisis and spurring Museveni to action:

> Museveni…learned of his country's AIDS problem from Cuba's president, Fidel Castro. At the September 1986 meeting of Non-Aligned Heads of State, Castro pulled Museveni aside. "[Castro]…told me that of the 60 soldiers we had sent to Cuba for training, 18 of them had the virus of AIDS. … [h]e was therefore worried that it may reflect what [level of prevalence] is in the [general] population." ('Uganda' 2005)[20]

The paternalistic role that Castro played with Museveni in Uganda, and in the Global South as a whole, has resounding symbolic impact on the consciousness of people on the African continent. For example, images of Castro frequently grace the covers of popular news magazines in Uganda, where he is hailed as a hero. This kind of news coverage is consistent with the perceptions Ugandan people express in everyday interactions with Cubans. For example, for several years Cubans served as professors at Gulu University in northern Uganda. Ugandans who encountered the Afro-Cuban professors were often confused because these men possessed African phenotypes but did not appear to be from Uganda. In an attempt to determine their origin, Ugandans consistently approached the *internacionalistas* by trying out various African tongues – Swahili, Arabic, and Lingala – assuming that these men hailed from some distant tribe in another part of Africa. When they eventually discovered that these men were Cuban, the Africans inevitably responded with a comment like, "Ah, Cuba! Fidel Castro has defeated the United States!"[21] The United States is undeniably seen by many in the Global South as the economic and political leader of the world; but in the eyes of many Africans, Cuba has sufficiently frustrated the United States since this mighty superpower has found itself unable to fully subdue Cuba and its Revolution. And sending Cuban athletes around the world – these impressive physical specimens – reinforces a notion of the symbolic strength of the Cuban Revolution, thereby building solidarity throughout the developing world. For these reasons, many Africans see Cuba as a heroic nation that subverts the effects of severe economic embargoes and stifling political pressure, disrupting a generalized notion of United States hegemony and domination of capitalist forces across the globe.

Internacionalistas and the Health of the "Socialist Body" of the Cuban State

Aside from building up socialism on the global level, *internacionalistas* serve to strengthen the "social body" by buttressing the political, economic and socio-symbolic health of the Cuban State, both regionally and domestically. *Internacionalistas* work in this capacity as state-to-state envoys via two categories: *mission* and *collaboration*. A *mission* is classified as a state-to-state relationship that is primarily ideological in nature, of which a prime example is the Cuban mission to Venezuela.[22] In concrete terms, this relationship between Cuba and Venezuela takes the form of an economic exchange. Cuba currently dispatches as many as 20,000 doctors and other professionals to fill gaps in Venezuelan State services, and Venezuela reciprocates with heavily subsidized oil provisions to fuel the Cuban economy.[23] While there is a clear economic aspect to this exchange of "doctors-for-oil," the Venezuela mission also acts as a symbolic anchor for a burgeoning leftist-political block in Latin America and as a show of solidarity among its core players: Hugo Chavez in Venezuela, Evo Morales in Bolivia, and Fidel/Raúl Castro in Cuba. On the other hand, a *collaboration*, while still ideological, is constructed primarily as a state-to-state contract that is based on a request for services from the host country.[24] Both mission and collaboration assignments generally last two years, including one month-long break every year to return to Cuba. Sometimes an *internacionalista* can earn a higher salary on a mission than a collaboration,[25] a recognition for "doing an important job," as one *internacionalista* put it. In addition to earning more money, on a mission, *internacionalistas* gain other symbolic perks. "Every-one wants to go on a mission," one *internacionalista* explained, "it gives pres-tige." Prior to embarking on a mission, *internacionalistas* must attend a training seminar that emphasizes the ideological tenets of the mission. The financial benefits of a mission versus those of a collaboration are offset by other regu-lations; although a mission emphasizes an ideological bent, a mission assign-ment cannot be renewed, presumably to give others the opportunity to serve in this capacity. On the other hand, while *internacionalistas* working within the context of a collaboration earn a lower net monthly salary, they have the pos-sibility to renew the contract for an additional two-year term, thereby afford-ing the *internacionalistas* on a collaboration the potential to earn more overall cash during their work period overseas.

The Cuban government imposes a very strict set of requirements to be-come an *internacionalista*,[26] including loyal ties to the Communist Party and/or evidence of fervent support for the Revolution. Each candidate for *internacio-nalista* work undergoes a background check, including an investigation into the individual's criminal record and interviews with neighbors, co-workers, and neighborhood officials in the *Comité de Defensa de la Revolución* (or CDR).[27] As an *internacionalista* coach in Botswana said, "They want to know if these are straight people or not...[they want] people with strong values." The officials who investigate prospective *internacionalistas* undoubtedly aim to send people overseas who perpetuate the notion that the Cuban Revolution is a shining success. However, it is undeniable that there is an overriding practical concern driving these investigations in that the Cuban State takes great pains to ensure that envoys are loyal to the Revolution *and* that they have family bonds strong enough to deter them from defecting from Cuba once overseas.

Aside from socio-ideological and symbolic aims, the Cuban State uses *internacionalistas* as a mechanism to mediate the health of its economic system, in particular to deal with foreign currency flows, lack of consumer goods, and underemployment.[28] The current economic crisis in Cuba has been compli-cated by failures in the centrally planned economy and by the U.S. embargo. Economically, Cuba also grapples with low levels of foreign currency inflow into the domestic economy and suffers from high numbers of professionals in the Cuban labor market. Furthermore, few goods enter the Cuban market, and there is not enough foreign currency in the economy to buy the goods that are available. By sending people abroad to earn money, Cubans are able to send foreign currency remittances *and* buy goods from overseas to bring home – everything from soap to computers. And while the Cuban Revolution has admirably succeeded in educating its citizenry, it also has led to a glut of overqualified and underemployed professionals. By sending workers outside on a large scale, the Cuban State also provides full employment for some of its redundant labor force. This process guts the labor market by sending some of its most capable workers to showcase the accomplishments of more than 50 years of Cuban socialism.[29] Cuban athletes in particular – along with doctors and musicians – are internationally "branded" and in high demand; thus sports professionals are regularly and eagerly requested as *internacionalistas* overseas. In this way, on a macro-economic level, athletes play key roles in bolstering the parasitical articulation of "socialist" Cuba with the "market-driven" global community. This effectively subsidizes the existence of less globally market-able professionals within Cuba who are unable to command foreign currency.

Some Cuban *internacionalistas* are undoubtedly motivated by a personal so-cialist ideology, but others are in no way interested in the role of proselytizing revolution. While many Western aid workers in Africa are blithely ignorant of their ideological function, Cuban *internacionalistas* are fully aware of their larger ideological role. Even if *internacionalistas* wanted to ignore that role, it would be difficult because the Cuban Embassy regularly sends socialist literature to them in the field to distribute. This literature includes books like *"Witness to the Miracle,"* a collection of testimonial accounts about *Operación Milagro,*[30] or copies of *Granma,* the official newspaper of *El Partido Comunista de Cuba.* Other common inserts in these packets are English-translated printouts of *Reflexiones del Comandante en Jefe,* the weekly musings of Fidel Castro.[31] Through *internacio-nalismo,* the Cuban State bolsters the ideological and structural health of the "socialist body" through agents who not only embody the ideals of socialism, but also play key economic and political roles for the State.

Internacionalistas and the "Health Of The Socialist Body" of Individuals in the Cuban Household

While some *internacionalistas* work in Africa in part to support the ideals of the Revolution, for an overwhelming majority, altruism and ideology have little to do with their desire to work there. They are in Africa or elsewhere to help their family households survive within the Cuban economy. As the fam-ily household is the basic political unit of the body politic, this is the aspect of the "socialist body" that most clearly reveals the everyday, lived practices of Cuban socialism and the contradictions between its practices and its ide-als. While working overseas facilitates the acquisition of foreign currency and goods necessary to maintain the household, it also launches other counter-revolutionary sentiments – consumer desires, temptation to defect, breaking of family bonds, and local animosity overseas – that act as deterrents of health for the "socialist body" at the local, State, and global levels.

Internacionalistas build up the health of the "socialist body" at the household level by earning foreign currency or by acquiring Western consumer goods in Africa at relatively low prices, either for personal use or to sell in Cuba at a profit. In Uganda, university contracts pay *internacionalistas* US$300 per month, of which US$100 is remitted to the Cuban Embassy (as a fee or a tax of sorts).[32] *Internacionalista* contracts generally include provisions for housing and sometimes food. Over the average two-year contract span, if an *internacionalista* saves as much money as possible, he or she can take US$2000-4000 back to Cuba at the end of their two-year stint. The drive for foreign currency not only has roots in macro-level state economy, it also is firmly grounded in daily household needs in post-Special Period Cuba (Eckstein 2004). By bringing consumer goods and cash back to Cuba, *internacionalistas* help their families to survive within the Cuban economy and have a direct effect on the physical health of Cubans. As household budgets become tighter and tighter, these overseas funds brought by *internacionalistas* are increasingly used to provide basic needs like food in Cuba.[33]

With some of this money, *internacionalistas* buy goods in Africa to sell at a profit in Cuba upon their return, thereby increasing their cash intake. With the remainder of their savings from those purchases, most Cubans build up a foreign currency reserve to buy things in stores in Cuba where items can only be purchased with "CUC," or Cuban Convertibles. CUC stores sell mainly foreign-made goods or goods made in Cuba for the export or tourist market, including soaps, foodstuffs (like rum, beer, chocolate, black pepper), diapers, electronics, etc.[34] While CUCs allow Cubans to purchase consumer goods, it is vastly cheaper for *internacionalistas* to buy those commodities in Africa and bring them home to Cuba once their contract is completed. In my experience, most *internacionalistas* in Uganda returned to Cuba with at least ten boxes of clothes to give to family or to sell: boxes full of jeans, camisole shirts, shoes, and underpants. Almost all *internacionalistas* also carried a laptop back to Cuba. Other items commonly coveted by *internacionalistas* are television sets and microwave ovens.[35] Aside from clothes and electronics, *internacionalistas* returned to Cuba with an odd assortment of other items. Many took hundreds of watch batteries back to Cuba with them, a favored item because they are small and light (therefore easily carried back), hard to acquire in Cuba, and able to be sold at a high profit. Soap, also a favorite article for many return-ing Cubans, is another small item that is an inexpensive but welcomed gift for friends, family, and neighbors who lack access to CUC (and therefore to quality soaps), or which can be sold in the informal market at a relatively high profit. The realities of lived socialism and *internacionalistas* travels can lead to particularly striking incidents.

One 250-lb muscled former wrestler and current sports professor at-tempted to carry hundreds of soaps to Cuba after several years of working as a coach in Botswana. When he arrived at Heathrow Airport in London carry-ing a bag full of soap, the authorities suspected him of possible drug smug-gling. One by one the authorities cut the soaps in half to search for drugs, but after splitting one-third of his haul they realized there was nothing embedded in the soap bars. For this inconvenience, the Heathrow Airport officials gave the *internacionalista* US$100. While for other travelers this destruction of prop-erty would be a major annoyance, for this *internacionalista* the setback was seen as a boon. When he returned to Cuba, he gave the cut-up soaps as gifts to his friends as he originally planned to do (since foreign soaps in any condition are a prized commodity), and he happily saved the additional US$100, a sum equal to about five months of salary as a professor in Cuba.[36]

Aside from aspiring to return to Cuba with a television or laptop, most *internacionalistas* in Uganda save money to buy a car. Cuba is a monopolistic car market, because only the Cuban State sells vehicles, and as a result strict restrictions and hurdles limit car purchases. Importation of automobiles into Cuba is prohibited, for example, so all cars must be purchased through the Cuban State system. There are a controlled number of vehicles on the roads; while this policy is good for traffic congestion, it is bad for consumer cost. For the gas-guzzling Soviet-era Lada that is commonly seen puttering on Cuban roads, one must pay US$3000-4000; for a classic 1950s American Chevrolet model the purchase price might be as high as US$6000.[37] Given these exor-bitant costs (relative to the salaries earned), the only way to legally own a car is to document that you have a sustained income of US$500 or more per month for a period of over two years and to obtain an official form signed by a designated supervisor to prove that the salary requirement is met. With this signed document *and* the US$3000-6000 for the purchase price, a Cuban is then qualified to legally register to purchase a car. Given the wages one is able to garner in Cuba – the average monthly salary for a Cuban worker is US$17[38] – it is impossible for most Cubans to sustain the income requirement or to save the money needed to buy a car unless a family member works overseas as an *internacionalista*.

Internacionalismo strengthens the health of the Cuban socialist body ideologically by equalizing access to foreign currency between socio-economic groups at the neighborhood level. Since the *Período Especial* (or "Special Period") commenced after the fall of the Soviet Union, the Cuban government has been opening up remittance streams as a necessary response to structural economic shifts. While this policy alleviated some of the most dire challenges of the Cuban economy, it also created disparities in wealth that threaten the most basic goal of the Revolution: social equality. For example, since many people who fled to the United States after the rise of Fidel Castro were white, upper-class urban dwellers, remittance streams tend to flow disproportionately towards white Cubans in the cities (Eckstein 2010). In addition, since the tourism sector tends to employ Cubans of lighter complexion, even tourism monies tend to flow towards sectors of the population that are considered to be "whiter." The exportation of *internacionalistas* around the globe balances out some of these socio-economic disparities within Cuba. Since the Revolution provides access to education and training to Cubans to all citizens, the ranks of professionals is more balanced in terms of race, class, gender, and geography. Therefore *internacionalismo* directs more foreign currency to black Cubans and rural dwellers, thereby engendering both physical and symbolic health for the Cuban socialist body at the state and neighborhood levels.

In spite of the equalizing effect that *internacionalismo* has on access to foreign currency streams, there are still incredibly challenging political, economic, and social restrictions in Cuba. As a result, many *internacionalistas* take advantage of their positionality overseas and secretly lay the foundations while overseas to relocate outside of Cuba.[39] A significant proportion of *internacionalistas* remain in the countries in which they serve. A few find ways to stay on legally, but most *internacionalistas* simply defect and cut ties with Cuba completely, except for sending remittances to their families. I was told by one of my informants that there is currently a community of defected Cubans in Botswana.[40] Another informant in Uganda reported to me in 2007 that of 71 *internacionalistas* that were sent all around the world from his university, 43 defected.[41]

Until very recently many Cubans defected to run a small business overseas – a venture that was almost completely impossible to conduct in Cuba on a legal basis until November 2011, when the government instituted new rules to permit some businesses to operate legally.[42] Laying the groundwork to launch a business while overseas can be risky. Even after undergoing the stringent process to become an *internacionalista*, Cuban expatriates' movements continue to be closely monitored and reported to the Cuban Embassy. Suspicious activities worthy of report include expressing "ambition" or having "a capitalist mentality" – accusations that can have dire economic and social ramifications for themselves *and* their families back in Cuba. In the past if branded as a possible threat to the Revolution, Cubans could be blocked from promotion in their professional life and even ostracized by their neighbors, co-workers, extended family, and larger community. This could make life extremely difficult since reliance on others is critical for one's daily survival in Cuba. Even though many of these repercussions have become less severe in an increasingly open socio-political environment in Cuba, fear still exists due to past reprisals. So, many *internacionalistas* act somewhat indifferent to money and goods in public, particularly in front of other Cubans – and *especially* in front of suspected informants. Because of the perception that they are possibly under constant surveillance in Cuba and overseas, many internacionalistas prefer far-flung assignments in parts of Africa, such as Uganda or Botswana, instead of Ethiopia or Venezuela due to the large populations of Cubans in these countries. In most parts of Africa, as one *internacionalista* told me, "I can live freely without everyone watching everything I do." While he was referring to the freedom to accumulate money and foreign goods, it was clear to me that while living abroad for a few years in an isolated, lonely situation was emotionally difficult, it also provided a much needed psychological break from constant surveillance and damaging gossip in Cuban Revolutionary society.

Due to the risks associated with trying to garner extra money or making plans to defect,[43] many *internacionalistas* try to secure an invitation to return to Africa as a contract worker with a host university or a private company. Private contracts allow Cubans to work legally in Africa with full freedom to travel back and forth between Africa and Cuba. Most *internacionalistas* are only allowed to secure these kinds of contracts after retirement age: 60 years old for men, and 55 years old for women. A chemistry professor from Santa Clara agreed to a contract job as a professor at Kyambogo University in Kampala after retiring at age 60. With this change in contract, his salary increased from US$300 to US$1200 per month, so he planned to stay on in Uganda for at least an additional five years to secure his familial needs into his old age. The price of this opportunity was formidable – several years away from his wife, daughters, and beloved five-year old granddaughter – an aspect of the *internacionalista* experience that is incredibly difficult for these intensely family-oriented Cubans. In Uganda, *internacionalistas* tried to save every bit of cash possible, spending money on nothing outside of basic necessities like food, but almost all the Cubans would budget money to call their families in Cuba every weekend. But in spite of these costly efforts to stay in touch, years of absence result in personal losses – children and grandchildren grow up, family members die, and so on. Another chemistry professor from Gulu University in Uganda mentioned another repercussion:

> This Revolution has done more to break families than anything – we are forced to spend years away from our families, and when we return, we find that our spouses have fallen in love with others - it is natural. You cannot be alone all the time. This is [our] life…

In addition to taking the emotional risk of leaving his family for many years, the professor at Kyambogo University in Kampala was also forced to relinquish his lifelong membership in the Communist Party. He admitted with deep regret, "I had to turn in my [Communist Party identification] card to take this job." This man was a true believer in socialist principles and the Revolution, and he spent many years as the president of his local CDR chapter, but in the end economic necessities required him to make this personal ideological sacrifice.

While *internacionalistas* are sent to fill gaps in services that host states are unable to provide, there are other reasons an institution in a developing nation might request *internacionalistas*. For many universities or sports teams, having an international scholar or coach gives prestige and credibility to the institution. Of all the options available on the international market, Cuban professionals are in great demand because they are highly skilled and they request very low salaries. An international competition-level Cuban sports coach is given a stipend of US$300-700 per month in an overseas posting, while an African coach earns a salary of US$2000 and European/American coaches command salaries of up to US$10,000. Of course, hiring a Cuban might mean displacing an African professional, either figuratively or literally. An already tenured African professor or coach might continue to remain in his/her job and command a salary; however, in close proximity to these highly trained Cuban professionals, their African counterparts are often overshadowed. In addition, Cuban professionals are seen by some as taking jobs away from African professionals. At Gulu University in Uganda, the chair of the Sports Science department initiated a campaign to get rid of the *internacionalistas*. He called the Cuban coaches into his office to announce that their contracts would not be renewed because according to him, they did not speak English. One of the *internacionalistas* responded, "But we are speaking English to you right now." "No," the chair responded as if not to acknowledge the *internacionalista*, "you do not speak English." This exchange was baffling to the Cubans, but later they learned from another university administrator that the Sports Science chair wanted to hire some of his own friends into the department, but was unable to do so while less expensive and more qualified Cubans remained on staff. For these reasons, many *internacionalistas* face political infighting with their African colleagues.

Conclusion

Sports professionals and coaches promote the health of the "socialist body" through everyday practice by building up the economic, political, social and symbolic well-being of global socialism, the Cuban state socialism, and quotidian life within Cuban households. While the ideological goals are clear and effective in many respects, the side effects of *internacionalismo* in everyday practice subvert the objective of a healthy "socialist body" in multiple ways, through defections, jealousy of local counterparts in Africa, splitting Cuban families, and cultivating consumer desire. An examination of the political, economic, social and ideological roles of *internacionalismo* not only demonstrates many strengths of the Cuban Revolution, but also highlights interesting contradictions that the Cuban State and its citizens face in the lived reality of building up the "socialist body" in the midst of overpowering capitalist forces and U.S. global hegemony. Through medical, academic, and sports *internacionalistas*, the larger Cuban revolution moves out of the symbolic realm and gives life to the global, state, and local "socialist body." Through sports in particular, the health of the physical "socialist body" is maintained and becomes a tool through which the symbolic "socialist body" is reified, thereby sustaining and buttressing what might otherwise be a dying Cuban Revolution. In that sense, the symbolic and physical manifestations of the Cuban "socialist body" are one and the same, and the health of one necessarily brings health to the other.

References

Abrams, P. (1988) 'Notes on the difficulty of studying the state', *Journal of Historical Sociology*, 1.1: 58-89.

'A map of Olympic medals' (2008, August 4) *The New York Times*. Online. Available HTTP: <http://www.nytimes.com/interactive/2008/08/04/

sports/olympics/20080804_MEDALCOUNT_MAP.html> (accessed 15 February 2013).

Anderson, B. (1991) *Imagined Communities: Reflections on the Origin and Spread of Nationalism*, London: Verso.

Arbena, J.L. (1988) *Sport and Society in Latin America: Diffusion, Dependency, and the Rise of Mass Culture*, Westport, CT: Greenwood Press.

Brotherton, P.S. (2008) "'We have to think like capitalists but continue being socialists'': medicalized subjectivities, emergent capital, and socialist entrepreneurs in post-Soviet Cuba', *American Ethnologist*, 35.2: 259-274.

Carter, T.F. (2008) *The Quality of Home Runs: Passion, Politics, and Language of Cuban Baseball*, Durham, NC and London: Duke University Press.

Eckstein, S. (2004) 'Dollarization and its discontents: remittances and the remaking of Cuba in the post-Soviet Era', *Comparative Politics*, 36.3: 313-330.

— (2010) 'Remittances and their unintended consequences in Cuba', *World Development*, 38.7: 1047–1055.

Feinsilver, J. (1989) 'Cuba as a "world medical power": the politics of symbolism', *Latin American Research Review*, 24.2: 1-34.

— (2010) 'Medical diplomacy: the international dimension of Cuba's health system', in A. Cabezas, I.N. Hernandez-Torres, S. Johnson and R. Lazo (eds.) *Una ventana a Cuba y los Estudios cubanos/A Window into Cuban and Cuban Studies*, San Juan (Puerto Rico): Ediciones Callejon.

Gleijeses, P. (2002) *Conflicting Missions: Havana, Washington, and Africa 1959-1976*, Chapel Hill, NC: University of North Carolina Press.

Guevara, E. (2000) *The African Dream: The Diaries of the Revolutionary War in the Congo*, London: Vintage Books.

Malkki, L. (1997) 'The rooting of peoples and the territorialization of national identity among scholars and refugees', in A. Gupta and J. Ferguson (eds.) *Culture, Power, Place: Explorations in Critical Anthropology*, Durham, NC: Duke University Press.

Miller, T., Lawrence, G., McKay, J. and Rowe, D. (2001) *Globalization and Sport: Playing the World*, London: Sage.

Ramonet, I. and Castro, F. (2009) *Fidel Castro: My Life: A Spoken Autobiography*, New York: Scribner.

Ryer, P. (2006) 'Between la yuma and Africa: locating the color of contemporary Cuba', doctoral dissertation, University of Chicago.

Scheper-Hughes, N. and Lock, M. (1987) 'The mindful body: a prolegomenon to future work in medical anthropology', *Medical*

Anthropology Quarterly, 1.1: 6-41.

Taussig, M. (1992) *The Nervous System*. New York: Routledge.

'Uganda: a Complicated Success Story' (2005) in Frontline: age of AIDS, *PBS*. Online. Available HTTP: <http://www.pbs.org/wgbh/pages/frontline/aids/countries/ug.html > (accessed 15 February 2013).

'Uganda AIDS Commission' (2012) Online. Available HTTP: <http://www.aidsuganda.org>.

Wilson, J. (1994) *Sport, Society, and the State: Playing by the Rules*, Detroit, MI: Wayne State University Press.

Endnotes

1   I stayed in contact with the original set of *internacionalistas* in Uganda by email and phone. Most of these people returned to Cuba in late 2008, but several were later reassigned overseas (for example, in Ecuador, Uganda, and Botswana). All *internacionalistas* are unnamed in this article for the purposes of confidentiality.

2   For example, the head coaches for the Botswana national track-and-field team and the Ugandan national boxing team are both Cubans. During the Olympics in 2008, the boxing coach told me that the Ugandan delegation was placed next door to the Cuban delegation at the Olympic Village, and he arranged training practice bouts between Cuban and Ugandan boxers.

3   It should be noted, however, that several of those Cuban doctors secured prominent, influential, and lucrative positions while serving as *internacionalistas* in Uganda. One case of a Cuban *internacionalista* doctor securing a contract position in Uganda is an interesting – and for Cuban internacionalistas in Uganda, legendary – story. In the late-1980s there was a high-ranking Ugandan military officer who had a mysterious disease. The Ugandan officer was sent to India for treatment, but was eventually sent back to Uganda to die as the Indian doctors could not help him. As a last resort, someone suggested that the officer go see the renowned Cuban doctor in Mbarara, in Western Uganda. The Cuban doctor examined the officer, and announced to him that he would not die if he took some necessary measures. The officer complied with the doctor's orders, and miraculously made a full recovery. This news reached the President of Uganda who was so impressed by this surprising outcome, he requested the Cuban doctor to stay on as the head doctor for the Ugandan Military. As an *internacionalista* the doctor knew that he was required to return to Cuba after his 2-year

term ended, so Ugandan President Museveni appealed directly to Fidel Castro to release the doctor from the *internacionalista* program. Upon Castro's approval, the doctor was then released to work legally in Uganda for a salary rumored to be between US$80,000-100,000. The doctor held that post for over 18 years, during which time he was granted dual Cuban-Ugandan citizenship. He used this foreign passport to leverage not only consultancy contracts in Europe and other parts of Africa, but also to sponsor his brother to leave Cuba and come to Uganda. His brother was a renowned Cuban film producer who worked on Sydney Pollack's film *Havana*. The former filmmaker abandoned his life's work in the movie business after years of fighting myriad creative restrictions and moved into coffee production in Uganda, later inviting his nephew in Cuba to work on the business. To every Cuban *internacionalista* in Uganda, this served as an almost mythical example of how opportunities can fall upon a Cuban when working overseas as an *internacionalista*.

4   One could argue that the promotion of baseball (and sports in general) helped to cast a particularly masculine tone to the Cuban revolution.

5   While I was in Cuba visiting returned *internacionalistas* in Havana, the household enthusiastically watched on one of the main television channels a woman's volleyball game at 8pm on a Thursday night – notable to me because this is a "prime time" commercial television slot in the U.S. that would never feature this type of low-profile sporting event.

6   In both medical and sports forums, success of Cuba's brand of socialism is broadcast to the rest of the globe, and in both cases the body plays a prominent role in demonstrating that success. The Cuban State aggressively pursues raising internationally measured health indicators to show how with fewer resources Cuban health outcomes are comparable to or superior to those of the United States.

7   Less than ten years after the revolution, in 1968 Cuba started to make a showing in the Olympic Games, earning four medals – as many medals as Sweden. By 1976, Cuba was ranked in the top ten of national medal holders – earning 13 medals – more than Britain. In 1992, Cuba won 31 medals (14 gold medals) – ranked fifth in the world ('A map' 2008).

8   As one author lamented: "the Olympics …once stood for the ideal of excellence, today they are a money-making opportunity and stand for greed." (Miller et al. 2001: 13)

9   Thanks to Tim Adams, who worked on the coaching staff for American football expansion teams in Europe, for raising my awareness about the importance of *cost* in the effective spread of sports into the global market.

10 In 2012, the Cuban coaching staff in Botswana brought the track and field divisions to the All-Africa Games and the athletes placed higher in competition than ever before.

11 Of course the "socialist body" in its political form is shaped through both the phenomenonally experienced and the social body; therefore these forms are inextricably linked and intertwined and can only be dissected as a heuristic exercise. Thanks to Jan Brunson and Nancy Burke for directing my attention to much of the literature on the bodily politics of sports.

12 Africa's role in Cuban *internacionalismo* will be explained further in the next section.

13 Che was particularly frustrated with Kabila. Cubans have deep sociocultural and symbolic ties to Africa. At least 60% of the Cuban population is classified as "black" or "mulatto" attesting to some form of African ancestry, and Cubans commonly attribute their unique culture of musical and dancing styles to their African heritage. In addition, there is a deeply entrenched and active spiritual practice of *Santería*, a religion that has its roots in the geographical area of modern-day Nigeria.

14 See *Conflicting Missions* by Piero Gleijeses (2002) for more. Paul Ryer (2006) records how the conflict in Angola in particular plays a key role in binding Cuba symbolically to Africa, citing Castro calling those "blood ties": "I mean this in two ways: by the blood of our ancestors and the blood we have shed together on the battlefield. ...We have fulfilled our elementary internationalist duty with Angola. (Castro 1981:107 from Ryer 2006:172)

15 Even though my research on Cuban *internacionalistas* in northern Uganda was a small part of my larger dissertation project on the effects of humanitarian aid in an African warzone, Cuba's role in Uganda emerged in conversations on many occasions. One notable occurrence was during an interview with one of the most notoriously violent former commanders of the Lord's Resistance Army (LRA), Kenneth Banya. Before Banya's tenure as an LRA rebel, he was one of most prominent fighter pilots in the Ugandan military. He recounted a story to me of meeting Fidel Castro during a reception in Uganda during the early 1970s during the Obote II regime, thus illustrating Cuban-Ugandan relationship prior to the Amin years.

16 Not unlike the support Cuba is giving to Mugabe at the moment, as was the case with Qaddafi before his death.

17 All children in Uganda are entitled to a free primary education, but critics allege this accomplishment is overstated; since there is not enough funding for this initiative, this requires some schools to force 200-300 students into one classroom.

18 Almost half of the national budget comes from donor aid money from Europe and the United States, in part to support these initiatives.

19 These figures were reported for 1992 – the earliest rates I could find data for – but the mid-to-late 1980s HIV/AIDS rates were either relatively similar or perhaps less than those in the early 1990s.

20 Conversation with Dr. Lorenzo Lopez, chief doctor for the UPDF (Ugandan military), June 2008. "Museveni and his government acted quickly, airing television and radio ads and speaking about HIV/AIDS at political rallies. They developed the now-famous ABC campaign, which stands for Abstain, Be Faithful, or use a Condom. 'I could not keep quiet on this,' Museveni says. 'There was no other option. Any other option would have been murder.'" ('Uganda' 2005).

21 In East African parlance the use of the word "defeated" does not imply that Cuba has conquered or crushed the United States. Instead, here the word "defeated" has a connotation closer to "frustrated" or "not being able to overcome." For example, a university student struggling with writing an essay might say that, "this paper is defeating me."

22 And secondarily, Bolivia.

23 This oil source is critical to Cuba's economic survival, especially in light of the 50+ year U.S. embargo. In the past, this oil subsidy was provided by the USSR, which received cheap sugar from Cuba in exchange for cheap oil. However, the fall of the Soviet Union, and the subsequent end of the oil flow from there, caused a near collapse of the Cuban economy. This caused a severe economic crisis known as the "Special Period." This emerging relationship with Hugo Chavez has helped Cuba to recover ever so slightly with renewed access to Venezuelan oil reserves.

24 For example, in 2009 the Botswana Government requested the services of several top Cuban athletic coaches to prepare them for international competition in the 2012 Olympics.

25 Usually a collaboration assignment pays about US$300 per month, whereas I have heard of mission assignments that pay as much as US$700 per month.

26 Especially for a mission.

27 CDR – the mechanism through which the Cuban revolution operates at the local level. The CDR is considered to be a non-governmental organization, but works in conjunction with the State Party apparatus. These are the neighborhood "block" level forms of governance. They organize the obligatory *trabajo voluntario*, and the CDR disseminates many goods and services at the ground level. For example, before my visit to Cuba in 2009, the Cuban government made an order for millions of energy-efficient Chinese refrigerators to replace

the old ones in Cuba (which were probably made in the USSR during the post-WWII era or even may have come from the United States and dated back to the 1950s). The CDR then was responsible for disseminating these refrigerators to the public at the local level and receiving the nominal fee for receipt of the refrigerators and remitting it back to the State. The CDR also acts as the "eyes-and-ears" of the Communist Party in Cuba on the ground, although there are also informants that exist in each neighborhood who operate in secret.

28 This utilitarian economic aspect of *internacionalista* work and how it is used as a mechanism to mediate the labor economic situation appears to be an obvious departure from Western aid mechanisms. Western international aid is assumed to be altruistic in nature and completely devoid of any useful purpose beyond "alleviating poverty." But this is a misconception. For example, the top food aid donor, the United States, uses food aid as a valve to release surpluses and thereby keep food commodity prices stable for producers. And Norway, the top aid donor as a percentage of GDP, uses cash aid as a way to relieve inflationary pressures caused by too many petro-dollars in the economy. In the same way, Cuba uses *internacionalistas* as a valve to release pressures on the labor market.

29 According to one informant, sending the best workers out of Cuba leaves those who "come to work drunk every day" in charge of responsibilities they would normally not be privy to. Although I have not read anything on this, one could speculate that the phenomenon of chronic alcoholism in Cuba coupled with siphoning off the most capable of the workforce to become *internacionalistas* might contribute in some way to Cuba's waning productivity and shrinking domestic economy.

30 *Operacion Milagro* refers to the 2004 Cuban mission trip that sent surgeons to Venezuela and Bolivia to perform thousands of eye surgeries to restore sight to people who either could not afford the surgery or did not have access to this service in some form or another.

31 These tend largely to be rants against the imperialists to the North (the United States) or declarations of the triumph of the Revolution (*Reflexiones* is also printed in *Granma*). I teased one of the *internacionalistas* calling these papers their "propaganda packets" from the embassy. He retorted by saying, "No, it's not propaganda...it's the truth."

32 In 2008, the requirement to submit cash to the Embassy was dropped, and Gulu University was allowed to raise their pay for *internacionalistas* to US$700 by reclassifying the position as a political "mission" (and thereby raising the salary amount to the level of their African counterparts). This US$700 salary is very uncommon though - US$300-400 per month is the norm for salaries paid to internacionalistas.

33   Difficulties in securing food that emerged after the Special Period are com-
      pounded by the 2008 global economic crash and the severe hurricanes that
      continue to pound the island, including multiple hurricanes in 2008 and, more
      recently, Hurricane Sandy in October 2012

34   There is a complicated dual-economy system in Cuba that is reflected in the
      circulation of two currencies in the Cuban market: Cuban pesos and CUC
      (Cuban *Convertibles*). A multiple-currency system is a common economic practice
      in a centrally planned economic system (the same kind of system existed in
      Communist China) in order to separate and shield the "internal" domestic,
      centrally planned economic system from the whims of the "external" global
      "free" market. Foreign goods can only be purchased in Cuba with CUC – which
      is equivalent (and convertible) to foreign currencies. In the past, 1CUC=US$1,
      but after 2000 in defiance of the U.S. embargo (and because of declining terms
      of trade with a depreciating U.S. dollar), Cuba shifted to a Euro-based economy
      where 1Euro=1CUC. As for the usage of the CUC versus the Cuban peso,
      there is essentially a list of commodities that can be purchased with CUC, and
      another list for Cuban pesos. In general, CUCs were designed for the tourist
      market or the export market, so goods and services that may be used for the
      tourist or export market (certain types of beer/rum, taxis, certain foods, etc.)
      are purchased with CUCs. Cuban pesos are generally used for basic needs and
      subsistence items used by the Cuban population, as well as other goods and
      services produced in Cuba *for* the Cuban internal market (e.g. other beer/rum,
      local taxis, certain foods, etc.). It should be noted that since Havana is the main
      tourist area in Cuba, CUC use is much more pervasive than elsewhere in the
      Cuba, for example, in Santa Clara where I conducted some of my research.

35   Cubans I spoke with hypothesized that the officials did not want Cuban citizens
      operating fast food businesses out of their homes, so they made microwaves
      illegal to minimize this possibility. During my visit to Cuba, I found that one of
      the former *internacionalistas* had a microwave in his house. I asked how this loyal
      Communist Party member and socialist devotee acquired this piece of contra-
      band equipment. "Oh, it is legal – I bought this in Mexico before 1988!" (I as-
      sumed this was the time when the ban on microwaves came into being). To deal
      with this restriction, one *internacionalista* in Uganda was able to make contact
      with a Cuban *aduana* (customs) official through whom he arranged bring one
      microwave oven into Cuba with no bureaucratic problems. Later, Raúl Castro
      came to power in early 2008 and enacted several policy reforms, including the
      legalization of microwaves (and cell phones).

36   The need to buy cheap consumer goods in bulk has forged an interesting

relationship in Uganda between the Chinese business people in Kampala and the Cubans. While *internacionalistas* do not buy stock in the same volume as a Ugandan store owner might, the Cubans are notable visitors to the Chinese shops in Kampala because they are some of the only foreigners that dare to venture into the bowels of the chaotic African market. The Chinese markets are the only places one can purchase packets of 12 shirts, 20 underpants, and hundreds of batteries in bulk at rock-bottom prices. Since most foreign expatriates in Uganda earn obscene amounts of cash as aid workers or multilateral/bilateral bureaucrats (relative to their local counterparts), there is no need for most of them to risk a trip to the African market to buy at a discount. But Cubans who are paid salaries comparable to (or below) their Ugandan counterparts are forced into areas of this African city where "regular" Ugandans shop and live – an interesting spatial aspect of the Cuban existence in Uganda.

37   The car model is not the main determinant of price since "branding" is not a main determinant of price in a non-capitalist market. Instead, utilitarian concerns are of greater concern in price, for example, the type of engine – gas vs. diesel (diesel engine is more expensive). As an aside, I estimate that about 40% of cars on Cuban roads are Soviet-era Ladas and 30% are 1950s U.S. cars (e.g. Chevys, etc.). The remainder of the vehicles includes new cars from Korea or Japan, horse carts, bicycles and bicycle-taxis, or Chinese buses.

38   My most illustrious colleague in Uganda was a dean in Cuba at his University in Santa Clara, and therefore he earned US$29 per month.

39   For example, butchering a cow can send a person to jail for 20 years. Also butter consumption is restricted (and illegal for most people), so bread vendors call out in the street *"Pan con apellido"* ("bread with a last name") to signify to their customers that contraband butter is included with the bread.

40   I have not personally been to Botswana, I cannot find statistics on Cuban defectors in Botswana, and therefore I cannot verify this statement – this is information from an *internacionalista* serving in Botswana, who told me about "many" defectors there during a phone conversation in 2010.

41   Conversation with an *internacionalista* informant during August 2007.

42   Before this ruling, almost everyone had a "small business" in Cuba (whether legal or underground), ranging from selling cigarettes from the front door, to cutting hair, to making shoes.

43   If a Cuban defects, he or she is unable to return to Cuba to visit family for seven years or more.

Section III

Global, Local, and Personal: Biomedical Practices and Interpretations

8

Conceiving Statistics: The Local Practice and Global Politics of Reproductive Health Care in Havana

Elise Andaya

On January 3, 2005, just days after the collection of the final statistics for 2004, the lead headline in *Granma*, one of Cuba's two daily state-run newspapers, screamed "Infant mortality 5.8!" More sedately, the headline of the other daily, *Juventud Rebelde*, clarified, "Cuba achieves a mortality rate of 5.8 for every 1,000 live births." Placing Cuba's accomplishment in a comparative framework, a chart of selected international statistics appeared next to an image of a smiling baby, demonstrating that Cuba's infant mortality rate was slightly higher than that of Canada (5/1,000), but lower than the United States (7/1,000), and far below those of its neighbors, Mexico (23/1,000), the Dominican Republic (29/1,000), and Haiti (76/1,000). Announcing that Cuba's infant mortality rate - then the lowest in its history[1] - placed it among the top 36 countries in the world, both newspapers went on to state:

> this undeniable achievement in the protection of the first of the human rights, that of health, especially that of women and children, was achieved in a country besieged and under embargo for more than four decades by the most powerful might in the world that, on the other hand, exhibited an infant mortality of seven [per 1,000 live births].

Cuba's superb reproductive health statistics are an achievement of which the government is justifiably proud, as they represent the culmination of the state's political will and the investment of huge financial and human capital. Cuba's infant mortality rates have fallen every year, even during the severe economic crisis of the Special Period. Despite massive shortages in food, basic goods, and medical supplies in the aftermath of the fall of the Soviet Union,[2] infant mortality rates continued their decline from 11 per 1,000 live births in 1989 to nine per 1,000 in 1995. In 1994, moreover, the Cuban Ministry of Public Health (MINSAP) established a low birth-weight clinic to address the increase in rates of low-weight babies caused by the drastic drop in Cubans' standard of living (Chomsky 2000). That a state under such economic duress as Cuba during the 1990s should continue to invest substantial resources into health care is testament to the government's commitment to maintaining - and improving upon - previous health gains.

Yet public health is not the only issue at stake. Cuban reproductive and infant health statistics resonate far beyond their medical or public health implications; they undergird the moral legitimacy of the revolutionary government both at home and abroad and are enlisted into the ideological battle to assert the continued vitality of the socialist agenda in a new world order. For the Cuban government, as the *Granma* article makes clear, reproductive health statistics are valuable arsenal in the fraught ideological exchange that characterizes U.S.-Cuban relations. Framed by the Cuban socialist discourse of health care as a human right, the health and survival of babies and pregnant women are central to state claims of the moral superiority of socialism over the inequality considered characteristic of human relations under capitalism (Castro 1967; Feinsilver 1993).

Cuba's superb health indicators also function as "ambassadors for socialism" (Feinsilver 1993) on a broader international stage and can be parlayed into other forms of international aid in the form of grants, loans, and collaborations. Cuban health officials and medical professionals pride themselves on the fact that the World Health Organization, the Pan American Health Organization, and UNICEF frequently cite Cuban reproductive health care as a model for other developing nations. Health statistics are thus key to Cuba's moral standing in these global forums, where health outcomes become an internationally accepted measurement of the good governance of states. When prominent international groups such as Human Rights Watch charged Cuba with lack of political freedoms, for example, Fidel Castro repeatedly pointed to Cuban health as evidence of the moral legitimacy of his government.

As part of its commitment to maintaining its place in the global stage, Cuba is also a signatory to the U.N.'s Millennium Development Goal, which aims to reduce infant mortality by two-thirds and maternal mortality by three-quarters by 2015. Given its already-low mortality rates, however, this may not prove an easy task. On World Health Day 2005, aptly themed "Every Mother and Child Counts," the director of the maternal-infant care division at MIN-SAP admitted, "It's not the same to lower a maternal mortality rate 75% from five [per 10,000 births, Cuba's 2004 maternal mortality rate] as it is from 100 or 150. But we are at a moment where we have the political will and the available personnel to make this happen."

Of course, reproductive health statistics depend not only on political will and medical personnel, but also on the women whose aggregate reproductive practices and birth outcomes become the health of the population. In this chapter, I present an overview of Cuban prenatal care as it played out in the family doctor clinics in which I conducted ethnographic fieldwork. Through analysis of patient-doctor interactions, I explore how doctors employ discourses of obstetric risk in their attempts to model ideal pregnant (socialist) patients. In the process, I also examine the coupling of nurturance and discipline in Cuban prenatal health policy as the Cuban state strives to produce desired subjects – both in utero and as patients/mothers – by exhorting both pregnant women and their doctors to bring desired maternal and fetal numbers, quite literally, into being. I argue that Cuban prenatal care is shaped not only by the international stakes tied to Cuban reproductive health statistics, but also by socialist ideologies with respect to ideal personhood and the relationship between the individual and the state.

This chapter is based on research conducted during 13 months of fieldwork in Havana 2004-05 and a month-long follow-up trip in 2009, during which period I explored shifting ideologies around gender, kinship, and reproduction in contemporary Cuba. As part of this research, I immersed myself in the busy ambit of my neighborhood family doctor clinic.[3] I observed prenatal and neonatal healthcare consultations twice a week for eight months, and accompanied doctors on home visits and on several overnight emergency shifts at the local polyclinic. These family doctors also introduced me to other medical personnel in their social and professional networks. Through these contacts, I observed weekly ultrasound consultations, as well as two ultrasound training sessions for doctors about to be sent to Venezuela; intermittent reproductive health consultations in two other family doctor clinics; counseling sessions with the neighborhood family psychologist; and I visited several neonatal intensive care facilities around Havana. Over the months of my fieldwork, I spent between five to 15 hours weekly with a core group of seven doctors and nurses. In addition, I was able to speak with and observe 15 other family doctors and specialists for periods spanning from several hours to several months. I took constant notes during these hundreds of hours of immersion in clinic life and wrote up extensive field notes at the end of each day. All names are pseudonyms and all translations are my own.

As a final introductory comment, I wish to draw attention to the difficulties in writing about Cuba in an often highly polemic international context. In this chapter, my intention is move away from simplifying and polarizing discourses that treat Cuban health care primarily as a vehicle through which to debate the mandate of the socialist government. The Cuban commitment to health care as social justice is, in my opinion, both laudable and self evident. This commitment is not just articulated on the state level; more particularly, it is borne out in the daily practices of Cuba's doctors who labor, often under demanding conditions, for the well-being of others. Rather, I examine the "politics of numbers" as local practices of prenatal care are inflected with state political goals, socialist ideologies about the relationship between the state and its (future) citizens, and global politics in contemporary Cuba.

Cuban Prenatal Health Care in a Family Doctor Clinic

The streets were already bustling by the time the family doctor clinic in which I observed prenatal and neonatal health consultations opened its doors to the waiting line of patients at 8:30 a.m. On the cracked and uneven sidewalks of this densely-populated Havana neighborhood, flower vendors set up their brightly colored stands while clusters of people on their way to work drank small shots of strong, sweet coffee sold from the windows of homes. Old men and women sat on the steps in front of their houses, chatting as they augmented their small state pension by selling the state-subsidized matches, cigarettes, and toothpaste provided through the ration system.

The clinic itself was unremarkable; except for the fact that the wall sported a now-faded and peeling revolutionary slogan – "Lies may go a long way, but in the end the truth prevails. *Viva Fidel!*" – the building was virtually indistinguishable from the houses and state-run businesses surrounding it. Its windowless and graying cinder brick walls were interrupted only by a narrow band below the roof, where latticed bricks permitted the circulation of both air and extremely high levels of street noise. Inside the sparsely furnished four-room clinic – a waiting room, two doctor's offices (one clearly a makeshift arrangement), and a small adjoining examination room – the mottled paint on its bare walls was adorned only by two UNICEF-sponsored AIDS education posters promoting condom use and a framed postcard claiming, "A Mother is the Purest Form of Love." Reflecting Cuba's chronic paper shortage since the fall of the Soviet Union, the educational brochures and reading material often found in doctors' waiting rooms were entirely absent. Collectively, the two doctors' offices boasted only five rickety wooden chairs, requiring a constant negotiation of seating as doctors, nurses, and patients passed from room to room.

In this rather unprepossessing building, with the aid of a rotating three-person team of nurses and the occasional visits of specialists, two family doctors – Dr. Janet Torres, an Afro-Cuban woman with a quick smile and easy manner, and Dr. Tatiana Medina, a vivacious brunette with an occasionally sharp tongue – cared for between 1,000 and 5,000[4] people living in the radius of a few blocks. Although the Family Doctor Program had originally provided one doctor and clinic per urban block in this area (an estimated 120 families), this zone was notoriously over-populated; apartments were frequently divided and subdivided into additional makeshift living spaces, meaning that Janet and Tatiana in fact supervised far more families. In addition, the massive mobilization of physicians to Venezuela, already well underway by the beginning of my fieldwork in mid-2004, had required the closure and consolidation of many family doctor clinics. The result was an increased patient load for the remaining doctors and greater pressure on clinic resources. Indeed, to an outsider, the clinic's atmosphere appeared informal and at times chaotic: patients wandered in and out to greet doctors and give them a small token of appreciation, interrupted consultations to ask a "quick favor" of the doctor (frequently a prescription for a family member unable to attend), answered the telephone, and located their own test results when doctors were otherwise occupied. During my fieldwork, a mandate from the Ministry of Public Health elicited a short-lived effort to close the office door during consultations to comply with WHO statements regarding patient privacy. Within a month, however, the doctors had been worn down by their patients' staunch resistance to this new policy and reverted to freely allowing people to hover in the office doorway as they awaited their turn.

In addition to providing routine health care for the neighborhood's inhabitants, family doctors were responsible for monitoring the health and well-being of pregnant women and their infants. Despite scarce resources, the doctors prided themselves on their attention to prenatal care and their adherence to international guidelines, and the following protocol was carefully followed for all the patients that I observed.[5] After a *captación*, or identification, of a pregnancy – usually through an ultrasound since urinary and blood tests for pregnancy are generally unavailable in the public health system[6] – family doctors make a preliminary prognosis. Based on evaluations of the mother's physical and emotional health, doctors categorize women as "to be followed" (for those with no major risk factors) or as "high risk" (for women with a history of difficult pregnancies or miscarriages, or who suffer from genetic conditions, diabetes, or high blood pressure). Following the WHO definition, Cuban health policy views health as a state of bio-social-psychological well-being, and the standard evaluation of a newly identified pregnancy includes a number of questions about whether the pregnancy was planned and/or desired, family relations, and household economy. Such questions are designed to give providers broader insight into their patient's social world and identify potential threats to a woman's health or to her ability to fully comply with pre-natal health guidelines. These questions are arguably less important in Cuba, where family doctors (usually also women, reflecting the feminized nature of family medicine both in Cuba and in many other countries) live in the community that they serve, maintain continuous relationships with their patients, and often have an intimate knowledge of their social and familial background. However, the interview questions do help to highlight pressures that may influence health outcomes or that constrain some women's ability to adhere to medically advised behavior.

For the first few months of pregnancy, women designated as low risk are seen once a month in the clinic for routine checkups of weight, measurements, and blood pressure. These evaluations are painstakingly handwritten in duplicate on patient records, one of which is maintained for clinic files and the other is kept by the pregnant woman herself. In addition to routine physical examinations, women receive a supplemented diet as part of their rations beginning from 14 weeks of pregnancy, their iron levels are tracked, urinary tests are given before every consultation to check for excess glucose (that might signal gestational debates) or protein (indicating a possible urinary tract infections or the development of preeclampsia), and both pregnant women and their partners are tested for HIV/AIDS every trimester (an emphasis that reflects Cuba's concern with maintaining its low HIV infection rates). Women also undergo a minimum of two ultrasounds to track fetal age and weight, as well as a more thorough ultrasound to identify congenital malformations.[7] In later stages of pregnancy, the number of clinic visits rises to twice a month, culminating in a weekly clinic visit in the final month. At 36 weeks of gestation, all women enjoy fully-paid prenatal leave and – as most working women are employed by the state – many women in jobs considered physically or emotionally taxing are given medical certificates that permit them to take leave even earlier in their pregnancy. Women with "high-risk" conditions attend clinic consultations twice a month throughout their pregnancy, as well as special consultations at the local hospital to track their specific condition. Those women who develop complications during their pregnancies may also be referred to an in-patient maternity home (*hogar maternal*) for further supervision, an aspect of prenatal care that I discuss in more detail later in this chapter.

In addition to being monitored by a family doctor and a gynecologist, women attend a consultation with a nutritionist, a dentist, and a geneticist, as well as with a family psychologist who evaluates their emotional preparation for their pregnancy, the quality of their familial relations, and impresses upon them the importance of breastfeeding. New and expectant mothers are also paid monthly home visits by a doctor or nurse to evaluate home and familial environments, perform routine health checks, and teach new parents or mothers-to-be about diet and hygienic practices. They look for signs of social problems or "dysfunctional families," such as poor living conditions, alcoholism, and malnutrition, which may pose risks for the health of mother and child.

This remarkable dedication of resources to biomedical prenatal care, particularly considering its marginal and underfunded status in many countries, emerges in large part from Cuba's revolutionary history and the historical commitments of its leadership. Cuban maternal and infant health policies are rooted in the socialist commitment to health care as a human right in which, as Julie Feinsilver (1993) has argued, the health of the individual is seen as a metaphor for the health of the nation. Within this, the well-being of pregnant women and children in particular is deeply symbolic. Arguing that global inequalities should not be reproduced through disparities in access to, or quality of, health care, Castro early rejected the "barefoot doctor" model of many developing nations that rely on systems of paramedics and community health workers. Rather, the state developed a physician-based model like those in the U.S. and Western Europe as a means of delivering basic health care. Such developments also impacted reproductive care. By the 1970s, midwifery had been marginalized as a remnant of Cuba's "backward" past, and hospital-based birth, which had previously been restricted to urban areas, had become routine in most parts of the country. Today, 99.9% of all births take place in hospital. The medicalization of reproduction in Cuba was thus intimately tied to the revolutionary drive toward national modernization, socialist development, and Fidel Castro's often-articulated aspiration for Cuba to become a "world medical power."

As in other socialist countries, state and medical intervention into reproduction in Cuba became a means to demonstrate socialist superiority and rationality through managing the "quality" of the population (Anagnost 1995; Greenhalgh 1994, 2003; Kligman 1998). The Cuban state argues that social and financial investments in prenatal care have significant rewards in the long term; healthy children require less curative medical care and possess greater potential for development, which leads to greater work and educational capacity. Since the state, despite thinly spread resources, assumes financial responsibility for the health of the individual from cradle to grave, it has a particular interest in preventing the birth of low-weight and health-compromised babies who have higher mortality rates and, if they do survive, may suffer considerable developmental and health problems throughout their lives. The eugenic possibilities of this approach are obvious: while Cuban guidelines for genetic counseling insist on a nondirective approach, some observers have suggested that abortion is frequently counseled for women with conditions that might lead to negative birth outcomes (Feinsilver 1993; Whiteford and Branch 2008).[8] At the same time, however, Cuban prenatal health policies stand in contrast to the "medical heroism" of market-driven health economies which tout their successes in saving severely health-compromised babies while ignoring the fact that women's inadequate access to prenatal care and nutrition may contribute to these outcomes, as well as the continued labor and financial outlay of the families of "survivors" as they care for children with ongoing health problems (Casper 1998; Landsman 1998; Rapp and Ginsburg 2001).

Cuban health policy thus explicitly ties biological and social reproduction together through the language of social costs and population management (Vidal et al. 1991). For the Cuban state, as I explore in the remainder of this chapter, the reduction of risk through the provision of prenatal care falls into a broader socialist history that aims to demonstrate the superiority of socialism through the (re)production of healthy and disciplined persons.

Discourses of risk and discipline in cultural context

One morning in the clinic, the visiting obstetrician-gynecologist (ob-gyn) invited me to sit in on a consultation with a young woman during her initial prenatal assessment for her first pregnancy. Although the family doctors assumed the bulk of the routine care for the pregnant women in their neighborhood, their progress was also monitored by an ob-gyn who rotated her visits among a number of family clinics. The specialist currently assigned to this clinic was Yaína, a lively and outspoken Afro-Cuban woman in her late thirties, a devout follower of Santería, a Cuban syncretic religion, who was scheduled to be present in the clinic about once a month, although I could in fact detect no predictable routine. (This impression was later confirmed by Janet, who noted that the constant departure of physicians on international missions and the subsequent shortage of specialists had disrupted the regularity of their visits and the continuity of patient care). As part of this routine examination, Yaína instructed her patient about the desirable prenatal diet: "NO: pizza, spaghetti, fried foods, butter, ham, mayonnaise, beer, pork, biscuits, cookies, bread . . ." Seeing her patient's dismay as she listed many of the most common foods in the Cuban diet, Yaína said, not unkindly,

> See? Now you've started! Now you have to be disciplined (*disciplinada*). "Disciplined" means you have to come to the consultations, follow your treatments, do what we tell you to do. It's about priorities.

The exercise of discipline as a means to minimize risk and optimize birth outcomes was a constant theme during prenatal consultations. Such strict recommendations – aiming to encourage pregnant women to avoid foods with excessive salt and fat, and to incorporate more fruits and vegetables into their diet – were given not necessarily in the belief that women would follow them precisely, but rather as a means of impressing upon them the seriousness of the endeavor on which they had embarked and the necessity of prioritizing the health of their pregnancies over other pressing concerns. Pregnant women were constantly urged to take responsibility for their pregnancy through vigilant self-discipline achieved through monitoring of diet, weight, blood pressure, and psychological well-being, as well as through diligent attendance at the many required consultations.

For clinic staff, discipline was a central characteristic of the ideal pregnant subject. Both doctors and nurses often praised "good" patients – those who faithfully attended all the obligatory consultations and maintained their weight and blood pressure between desired norms – by describing them as "disciplined" (*disciplinadas*). Discourses of discipline were intimately intertwined with exhortations to comply with medical authority: one morning, I sat in on a consultation between Yaína and a woman in her early thirties, who began the session by telling the ob-gyn about her complicated reproductive history and family life. Despite this litany of stresses and difficulties, when Yaína reviewed her medical record, she said, "You have to be very disciplined because you've already had one premature birth and 12 abortions.[9] There are problems you can avoid with discipline and some you can't. . .We're here to help you, but if you aren't disciplined the fault is your own." In this commentary, Yaína draws on the language of risk to argue that the patient must strictly monitor her progress and adhere even more closely to medical directive – be a disciplined patient – in order to optimize her birth outcomes. Any issue that emerged from her failure of discipline would be hers alone to bear. The connection between discipline, compliance, and the production of desired medical sub-jects was so strongly linked in medical discourse that it was even imputed to fetuses. On more than one occasion, the doctor measuring a fetus's image dur-ing an ultrasound would a comment off-handedly to me, "This baby is really disciplined - it's not difficult at all to get it in the right positions [to measure its head, abdomen, and femur length]." Such personifications of the visualized fetus, as scholars of reproductive technologies have noted in other contexts (see, for example, Mitchell and Georges 1997), are often highly revealing of desired cultural attributes (a big bouncing boy, a pretty girl).

In contrast to "disciplined" clinic patients, women who failed to adopt the risk-reducing measures advised by state and international institutions were viewed as lacking in discipline. Yurielis was a boisterous young woman easily recognizable for her rambunctious laugh, heavy cat-eyed makeup, and tousled, shoulder-length auburn hair. Although she had moved out of the neighbor-hood to live with her partner's family prior to her pregnancy, she had neglected to file the paperwork needed to formalize her change of residence and there-fore traveled over an hour by bus to attend consultations at the clinic to which she was still legally assigned. She was known for her casual attitude towards prenatal health guidelines, and the clinic staff constantly traded stories about her non-appearance at external consultations or her frequent admissions that she had forgotten to take the prenatal vitamins provided free of charge to every pregnant woman. In general, the doctors treated her with a mixture of exasperation and affection, their annoyance at her behavior tempered by her breezy good nature. One morning, however, I entered one of the consultation rooms to see Tatiana angrily castigating Yurielis for yet one more infraction, this time another failure to keep a specialist appointment intended to track her low iron levels. After Yurielis departed, once again promising to attend, Tatiana vented her frustration loudly to Mayda and the gathered nurses, "The mothers here are so undisciplined . . . Yurielis was supposed to have gone to that anemia consultation two months ago! I mean, if women can't even be disciplined and get themselves to their consultations, how can they possibly expect to be good mothers?"

Tatiana's judgment of Yurielis as "undisciplined" stemmed from her apparent refusal to follow doctors' recommendations, thus potentially putting at risk her own health and that of her future child. The lack of discipline suggested by this medical noncompliance, for Tatiana, not only made Yurielis a "bad" patient, but also reflected poorly on her preparation and commitment to mothering in general. (When I talked to her privately, Yurielis, by contrast, drew on her embodied sense of her pregnancy to make her own health assessments, noting, "I'm young, I'm healthy, I feel great, my weight is good, my blood pressure is good. Sometimes the bus doesn't come, or I can't get anyone to bring me an hour into the clinic. What's the big deal if I miss a few tests?")

Moral discourses of risk and discipline, of course, figure equally prominently in prenatal care in other, non-socialist countries. Anthropological research has documented the globalization of biomedical intervention into processes of reproduction (Browner and Sargent 2011; Gammeltoft 2007a; Georges 2008; Martin 1987; Van Hollen 2003). The concept of risk is central to this reproductive culture, and a key characteristic of biomedical reproduction is the proliferation of techniques and technologies for the identification, calculation, and management of risk. Paradoxically, far from eradicating risk, the increasing medicalization of reproduction has made risk omnipresent even in apparently "normal" pregnancies, requiring constant vigilance and oversight to ensure desirable outcomes (Taylor 1998).

In its attention to maternal-fetal health, Cuban prenatal care thus participates in this central contradiction: at the same time as it promises safer pregnancies and more control over birth outcomes, it also contributes to the construction of pregnancy as a pathological or risky condition that demands constant medical oversight. The elaboration of risk in Cuban obstetric care, as in many other countries with biomedical reproductive health care, has been accompanied by a fundamental shift in the locus of power from the embodied knowledge of women to a reliance on the authoritative knowledge of doctors and technology (Browner and Press 1996; Davis-Floyd and Sargent 1997; Duden 1993; Georges 2008; Jordan 1997; Martin 1987; Rapp 1999; Rothman 1989). As Laury Oaks (2001: 85) argues in her discussion of U.S. public health campaigns targeting women who smoke during pregnancy, risk is:

> a social construction that highlights the authority that health professionals bring to their relationships with clients and that forms social expectations about what constitutes responsible, or moral, living. Risk assessment is a disciplinary technology in that it establishes norms, perpetuates reliance on

medical experts, and seeks to motivate individuals to attempt to control risk.

In prenatal care, pregnant women are subjected to ever-expanding regimes of medical enumeration and surveillance – which Michel Foucault (1977) called discipline – through which their bodily fluctuations are recorded, tabulated, and analyzed in order to detect and minimize obstetric risk. Through this process of constant observation, institutions like the medical clinic seek to produce subjects who have internalized this disciplinary imperative and who conform to desired practices without the need for force. Discourses around risk thus form part of a moral landscape in which "good" patients proactively engage in behaviors thought to minimize risk. Discipline in this sense (and as I use it) therefore refers not to its common usage as a synonym for punishment, but rather to this productive nexus between governance and subjecthood.

Yet these apparently similar discourses of risk and discipline are also informed by core cultural values. Research in the U.S. and in the Britain has highlighted how definitions of risk associated with reproductive technologies tend to engage dominant values of individual autonomy and choice (Franklin 1997; Rapp 1999). Discourses of (self)discipline index a worldview in which each individual is believed to be endowed with the freedom and the ability to make the "right" choices for optimal health and is thus ultimately responsible for his or her well-being (Rose 2001).

In Cuba, by contrast, discipline is a key political and moral value that describes not only technologies of the self, but also the relationship of the individual to society. As in other socialist countries, individuals are envisioned "as organically related social beings" (Gammeltoft 2007b). State media, classrooms, and billboards constantly seek to forge desired subjects through intervening in everyday practices, and the image of disciplined socialist citizens - productive workers, dutiful parents, and fervent patriots - tirelessly working together for the development of Cuban socialist society emerges constantly in state rhetoric. The concept of discipline is thus central to broader ideologies of socialist governance and socialist subjecthood, in particular, the effort to produce *el hombre nuevo* (the new man). In this context, the accusation of indiscipline raised by Tatiana references a wider judgment about appropriate personhood and socialist citizenship than medical compliance alone. As suggested by her slippage between her characterization of Yurielis and her condemnation of an entire neighborhood ("the mothers here are so undisciplined"), the concept of indiscipline in prenatal care was also laden with class and, to a lesser extent, racial connotations. Clinic staff, shaking their heads over an altercation in the street or stories of a fight in the neighborhood the previous night, would frequently declare, "What can you expect? This neighborhood is undisciplined." Although simply pursing their lips in the direction of the street when I asked what they meant by this, it is impossible to separate this commentary from the neighborhood's reputation for poverty, disorderly conduct, underemployment and (despite its racially-mixed composition), as a home for *negros*, or Afro-Cubans.

In prenatal care, the ideal of the disciplined, compliant, and risk-reducing pregnant patient is deeply intertwined with the broader socialist national utopia of productive and disciplined socialist citizens who work together for the collective good. (Indeed, according to orthodox socialism, all citizens were expected to contribute to the building of society according to their abilities. One of the capacities unique to women, as Castro early declared, was their labor as the bearers of children.) By contrast with the U.S., for example, the Cuban concept of discipline is embedded in a cultural context that emphasizes the role and responsibility of the broader society and community in helping women achieve a healthy pregnancy. State rhetoric positions doctors as part of the ideological vanguard whose task it is to uplift even the most unwilling of the masses, and clinic staff are required to follow up on women who miss prenatal or newborn appointments through telephone calls or home visits. Members of mass organizations such as the Committees for the Defense of the Revolution (CDR) and the Federation of Cuban Women (FMC) may also be called upon to facilitate the attendance of women encountering difficulties in attending prenatal consultations (providing childcare or transportation, for example) or to press reluctant women into compliance. With the weakening of these organizations in some neighborhoods, such tactics may not be employed as often as previously. However, the doctors with whom I worked did indicate that these resources were still available to them if necessary. Some observers have criticized these policies as excessive and medically unnecessary or as further evidence of the punitive functions of the state surveillance apparatus (Crabb 2001). Indisputably, and particularly given the weight placed on reproductive health statistics in the Cuban context, such efforts underscore women's further insertion into biomedical disciplinary regimes of enumeration and surveillance during the prenatal and postpartum periods as opposed to other moments in their lifecycle. Yet at the same time, such practices cannot be disentangled from a larger, and long-standing, cultural and political emphasis on the integration of the individual with the community.

Of course, these social and political ideologies did not always go uncontested by clinic staff, many of whom privately asserted the primacy of the patient's individual responsibility for her health. On one occasion, a student doctor interning in the clinic commented on one pregnant patient who had missed a prenatal appointment and seemed unresponsive to the doctor's comments and advice. Chiming in, Janet said:

> Well, it's because people here have the bad habit of thinking that it's the doctor's responsibility, that it's the doctors who should be in the streets, climbing the stairs to their apartments, reminding them [of appointments]. But this is *your* pregnancy, *yours*, and the one who has to live with this child is you, not me. But they think that the responsibility should fall on the doctor.

Embedded in this commentary is a medical morality based on assumptions of individual autonomy that coexists in tension with official discourses about socialist medicine and society. Rejecting older models of socialist medical practice in favor of a discourse of individual responsibility, she suggests that women, as future mothers and caregivers, must shoulder a greater portion of the responsibility for the management of prenatal risk.

Such discussions underscore an ongoing debate, especially among professionals, about the extent to which Cuban socialism has fostered an unhealthy dependence on paternalist policies (Andaya 2009). They also highlight the changing context of medical practice in contemporary Cuba, where the massive mobilization of physicians and medical support staff on international missions, particularly to Venezuela, has left remaining doctors feeling overwhelmed with increased patients and responsibilities. More broadly, as I address in the following section, they speak to the tensions, even in the Cuban context, between social and individual responsibility for health and risk management, as well as the sometimes conflicting roles and obligations of doctors in ensuring positive health outcomes.

The Janus-Faced Nature of Nurturance and Discipline in Prenatal Care

One hot and humid early morning in April, the bright new polyclinic building where I observed ultrasound consultations once a week seemed particularly cool and refreshing, its white and glass walls, tiled interior courtyard, and well watered plants offering a welcoming contrast from the dust and noise from the cobbled street outside. Inside the ultrasound room, Dr. Marisa Sánchez, a family doctor who was filling in for an ultrasound technician who had been sent on an international mission, worked in near silence on an aging ultrasound machine donated some years prior when an unknown overseas hospital updated its technology. Periodically, she called out measurements and diagnoses to the nurse, who inscribed them manually on the patient's health booklet.

On this particular morning, the routine was momentarily broken when the door opened and a teenage girl peered shyly into the dim room. Blocking her entrance, the nurse asked for her referral, but Marisa, who recognized her as a patient from her own clinic, broke into a rare smile and affectionately responded for her, "She doesn't need a referral, I dream about this young girl, this *muchachita*, I see her face when I eat my beans!" We laughed, and Marisa went on, "It's true, because in my 16 years of work I've never had a fetal or maternal death, nor a low-weight nor a growth-restricted baby. But it seems like this girl is going to stain my record, she's going to bring me a low-weight [baby, *un bajo peso*] and lose me my case of beer [that is frequently given as an end-of-year bonus]."

A few minutes later, as she measured the shadowy images of the fetus resolving in the ultrasound monitor, Marisa shook her head, saying, "In my opinion, it's still low weight . . . It has the head of a fetus of 33 weeks, not your 36 weeks." Addressing me directly, she went on, "For me, the most important thing is that she gets to 40 weeks . . . I want her to have a child of 5.6 lbs, just 5.6 lbs [0.1 lb over the internationally-accepted designation of low birthweight]. I've already admitted her to the maternity home so many times, made her stay in the hospital for days at a time, all so that she doesn't climb the stairs, so that she rests, and still . . ." The nurse looked at her admiringly, exclaiming, "You did all that?" and then, shaking her head at the young woman in mock disapproval, commented, "How lucky you are to have a doctor that cares so much!"

Marisa's comment and the nurse's response throw into relief the Janus-faced nature of nurturance and discipline in Cuban prenatal care. As in other contexts of biomedical care, doctors employ discourses of risk and discipline in attempt to mold women's behavior and improve birth outcomes. Yet, by contrast with most other countries, women who do not attain or maintain desired norms in pregnancy and are thus considered at elevated risk can be referred to hospitals or to in-patient maternity homes in the hope of more overtly influencing birth outcomes. Inaugurated in the early 1970s, maternity homes first primarily served rural women who often lived at great distances from the nearest maternity hospital. Since Cuba continues to insist on hospital births as the safest for mother and child, rural women nearing their due date are sent to a maternity home to ensure a swift transfer to a hospital at the onset of labor. Maternity homes are now ubiquitous throughout both urban and rural areas and function to monitor the health of pregnancies that may produce a low birth-weight baby or a fetal, neonatal, or maternal death. They provide enriched diets high in dairy, fruit and vegetables, and meats - foods often scarce in the diet of the average Cuban - as well as round-the-clock care from a staff of specialists who follow women's physical progress and teach classes on breastfeeding and early infant childcare. In this atmosphere, women are supposed to rest and relax, free from the demands of family and domestic chores.

These concerns about the health impact of women's domestic duties were not insignificant. In the family doctor clinic, much of the discussion of women with "problem" pregnancies focused on women whose health concerns were attributed to familial pressures. While most women reported being "spoiled" by their families during pregnancy (no doubt a consequence of Cuba's extremely low birth rate – in 2011, about 1.5 children per woman – and the fact that many women would carry only one pregnancy to term in their lifetime), this was not the case for all. A complaint frequently voiced by doctors was that the demands of their patients' families – particularly their partner – that women continue to fulfill domestic obligations related to cooking, provisioning, and cleaning contributed to insufficient weight gain, high blood pressure, and threats of premature birth due to the constant lifting of heavy objects. In such cases, doctors would often recommend bed-rest as a way of both relieving women from her domestic duties and of averting a preterm birth that could result in a newborn health complications or fatality. When they suspected that women were continuing to work excessively around the home, Tatiana or Janet would often telephone the patient's family directly from the clinic to reprimand them for not enforcing the recommended bed-rest. The women themselves received these blunt interventions differently; while some commented that perhaps now their husbands would listen, others were visibly uncomfortable during the telephone call, and the doctors themselves were often uncertain about the effectiveness of these strategies.

If women's obstetric health indicators failed to improve after these first-level interventions, doctors brought up the possibility of admittance into a maternity home. Cuban news programming abounds with television spots highlighting various maternity homes around the country, during which women invariably praise the high quality of food and care that they have received. Yet this portrayal of maternity homes differed markedly from my own observations. In the clinic, admittance into a maternity home was usually framed as a good-natured, but serious, threat should the woman fail to adequately discipline herself. Admonishments like that given by Tatiana to one patient – "You've been undisciplined, but that's enough already, because you have hypertension. If you're not disciplined, I'll put you in the maternity home because I don't know what you're eating" – were not unusual around the clinic. On another occasion, I sat with Janet as she examined a slender young woman in her second trimester whose weight gain fell below the desired norms. Frowning, she asked, "Are you eating well?" When the woman responded in the affirmative, Janet shook her head again, slightly disbelievingly, and declared her intention to refer her to a maternity home. Met with the woman's violent protestations, Janet relented, but declared, "I'll give you two weeks to gain the weight, otherwise you're going straight to the maternity home." After the patient departed, Janet told me she believed that the woman was sharing many of the additional rations of her prenatal diet – particularly the dairy allotments – with her child who, having just turned seven years old, had recently become ineligible for the additional dairy rations provided to young children. While other families supplemented these decreased rations with products purchased outside the *libreto* (the ration book), the very tight domestic economy of this particular family might be limiting their ability to sufficiently nourish all their members. In a maternity home, Janet hoped, the woman's food intake would be monitored and supplemented to increase her weight gain.

As these examples suggest, patients almost invariably reacted to the mention of maternity homes with expressions of dismay as they attempted to cajole the doctor into giving them a little more time to achieve the desired state. As many women were entered into maternity homes multiple times during their pregnancy, local institutions were frequently full. Women would therefore occasionally "escape" (*escapar*, Janet's phrasing) for just a few days or, in the least desirable scenario for all involved, were sent to maternity homes or hospitals outside the immediate area. Yet when asked directly, women generally understood these referrals were an expression of the doctor's concern for the health of their pregnancy – a factor that was ultimately more important than their dislike for a facility that was "not comfortable like home," or their anxiety about the impact their temporary absence would have on their family. The doctors themselves at times expressed sympathy for women's aversion; Janet agreed with one patient that conditions at some maternity homes were *pesado*, or depressing, since women often had to share a room and family was not able to spend the night, there was only one telephone available for use – significant in a country in which cell phones are still relatively limited – and the food, while healthy, was not actually *good.* At the same time, she too viewed maternity homes as an important resource in providing women with the additional medical attention to ensure healthy pregnancies.

In embedding pregnant women even more deeply into biomedical regimens of enumeration and oversight in the effort to improve maternal/infant health, maternity homes make explicit the twinning of nurturance and discipline in Cuban prenatal care. Some readers may find this overt medical surveillance over pregnant women discomfiting, representing the over-reaching of the state into the private affairs of its citizens or the ever-increasing penetration of doctors and technology into processes of reproduction. While sympathetic to such reservations, I suggest that such policies must also be judged within the Cuban context, in which the state has explicitly assumed responsibility for the health of population and in which the balance between the rights and responsibilities of the individual and the state has been struck differently than is the case in market-driven health economies (Whiteford and Branch 2008). Cuba's policies could also be favorably compared to those of many developing countries where the paucity of resources devoted to state-sponsored prenatal health care leads to high rates of maternal and infant mortality. Closer to home, maternity homes should be contrasted with the punitive policies of many U.S. states, where women can be incarcerated for fetal endangerment, yet little help is provided in the form of drug rehabilitation or prenatal/maternal support to help at-risk women.

Such policy distinctions underscore the Cuban emphasis on social responsibility for prenatal health, as well as the different valences that discourses and practices of nurturance and discipline assume in particular local contexts. Cuban health care, however, is also notable in that state oversight encompasses not only the population at large, but also the doctors who are considered responsible for the health of their communities. As suggested in the previous account by Marisa's half-joking concern that the birth of a low-weight baby might cause the loss of her bonus case of beer, doctors' own performance is monitored by the Ministry of Public Health, and they can be held accountable for undesirable birth outcomes.

The position of family doctors as gatekeepers between the health of individuals and the well-being of national statistics was underscored by one interaction in the clinic. Tatiana had taken a month-long leave and Janet was struggling to absorb her colleague's patients as well as her own. I looked up from my notes of the previous consultation to see Janet frowning at the sight of her next patient, a very thin young woman in a relatively advanced stage of pregnancy who handed over a plastic bag that contained her clinical history from previous prenatal visits and her most recent ultrasound report. Absorbed in the hand-written notes, Janet noted, "The ultrasound says that the baby's weight is normal, on the low side, but normal. But you're very thin, how much have you gained?" She made a quick calculation and then looked up with horror, exclaiming, "You've only gained 2.5 kg [about 5.5 lbs] in your whole pregnancy? This can't be!" Aghast, she called Tatiana, exclaiming, "This is awful! I have to admit her to a maternity home, because this can't be - the home right here, a hospital, it doesn't matter where, because she's only gained 2.5 kg in her whole pregnancy and she was thin, maybe malnourished, to start with!"

Attracted by the scandal, two of the nurses walked in. Assimilating the situation, one added reprovingly, "You have to rest, my dear, because your child is going to be born low birth-weight." Still flustered by her discovery, Janet added, "Yes, you have to have complete rest, I have to admit you at least until your weight stabilizes, you have to get all the tests done - this is awful!" The young woman's face registered some displeasure at this prospect, but when she suggested that she could just visit the maternity home on an outpatient basis, Janet quietly informed her that that was unlikely to be permitted. As she left, a waiting patient commented on her thinness. Shaking her head, Janet concurred:

> Yes, and if she gives birth to a low-weight [baby], it's a huge mess for us, because they send [the results] through the hospital, through MINSAP, they analyze everything, what happened, why a low-weight [baby] was born . . . [because if the baby is born low birth-weight] it will raise the rate of low-weights in the country, and it could raise the [neonatal] mortality rate as well. Just this year, in this polyclinic zone we've had two neonatal deaths, one from congenital malformations and the other, I don't remember. . . anyway, they counted them as two neonatal deaths. And with these two deaths, because we have so few births, the rate in this municipality is now so high that it raises the rate in the whole country, just with this municipality. We've got problems! (*Estamos en candela!*)

Situating this potentially problematic pregnancy within the context of the neighborhood's troubling recent birth outcomes, Janet's immediate concern was not "simply" the deaths of two infants, but that those deaths were categorized as neonatal mortalities (rather than fetal deaths) and would thus be entered into the reproductive health statistics for the municipality. In this context, the prospect of a child who might be born low birth weight, and whose mother's own low weight gain had somehow slipped through the cracks, would reflect poorly on the doctors' own performance. Should the child not survive, add another black mark to the record of a municipality that was already responsible for raising the national infant mortality rate. Janet's realization of her patient's poor weight gain thus triggered a cascade of interventions aiming at averting the birth of a baby whose potentially compromised health prospects might entail prolonged suffering or hardship for the child and their family, reduce their well-being and productivity, and blemish Cuba's national reproductive health statistics.

In both providing the nurturance required to ensure the birth of healthy babies and urging women to comply with prescribed prenatal practices, family doctors serve as the gatekeepers between the well-being of individual women and that of national reproductive statistics. This is not to claim that medical care in Cuba is driven primarily by doctors' awareness of state oversight. Medical professionals are genuinely concerned about their patients' welfare - these are, after all, neighbors and often friends. Yet they are also positioned as the "moral representatives" of the state in the juncture where individual women's health practices become internationally touted reproductive health statistics. Just as pregnant women are nurtured and disciplined within biomedical regimes of knowledge, doctors are also enmeshed in practices and processes of surveillance and enumeration as the socialist state strives to improve its reproductive health indices through its policies and practices of prenatal health care.

Slipping between her anxiety for the welfare of her patient and that of national reproductive health statistics, Janet's dense commentary neatly illustrates the simultaneous functioning of reproductive health statistics in Cuban prenatal health care as technologies of scientific knowledge, of government administration, and of symbolic representation (Urla 1993). In the nexus of state political goals, the international circulation of reproductive health statistics as a measurement of state morality and legitimacy, and local practices of prenatal care, discourses and practices around risk, nurturance, and discipline underscore the anxieties around the reproduction of both healthy children and the international moral standing of the Cuban government.

Conclusion

Scholars of reproduction have pointed to the emergence of a globalizing biomedical culture of reproduction in which ideologies of risk management, and its attendant forms of discipline, are central pillars. While underscoring broad similarities, this research has also highlighted the necessity of attention to what is not shared in this emergent global culture; that is, the distinct ways that discourses and practices of risk and nurturance are framed within locally specific configurations of disciplinary power, as well as how women may enable, embody, or contest these values in their reproductive practices.

In this overview of prenatal care in Cuba, I have pointed to the ways that both Cuban women and their doctors are positioned within the Janus-faced discourse of discipline and nurturance in state, international, and medical regimes. Close examination of the discourses of risk and discipline at the most local of levels – the neighborhood family doctor clinic – casts a bright light on Cuban prenatal health policy as it is interpreted and experienced by medical staff and their patients. Drawing together this ground-level analysis with an awareness of the international stakes of reproductive health statistics for the Cuban state also illuminates the relationship between the state and its citizens as the former attempts to influence the production of desired subjects - both in utero and as patients/mothers – through discourses and practices of discipline and nurturance that encompass both pregnant women and their doctors. With the state's recent abandonment of the previously espoused ideal of one doctor per urban block and the subsequent closure and consolidation of many family doctor clinics, many Cubans' experience of healthcare is changing. However, given the importance of reproductive health statistics to the Cuban state, both laypeople and Cuban health professionals with whom I spoke agreed that the provision of prenatal health care will remain a priority. This emphasis further underscores how local prenatal health care practices in Cuba are deeply informed by state goals, global politics, and socialist values, with profound ramifications not just for the reproduction of individuals, but for the reproduction of socialist state morality in the international arena.

References

Anagnost, A. (1995) 'A surfeit of bodies: population and the rationality of the state in Post-Mao China,' in F. Ginsburg and R. Rapp (eds.) *Conceiving the New World Order: The Global Politics of Reproduction*, Berkeley, CA:

University of California Press.

Andaya, E. (2009) 'The gift of health: socialist medical practice and shifting moral and material economies in post-Soviet Cuba', *Medical Anthropology Quarterly*, 23.4: 357-374.

Browner, C.H. and Press, N. (1996) 'The production of authoritative knowledge in American prenatal care', *Medical Anthropology Quarterly*, 10.2: 141-156.

Browner, C. and Sargent, C. (eds) (2011) *Reproduction, Globalization, and the State*, Durham, NC: Duke University Press.

Casper, M. J. (1998) *The Making of the Unborn Patient: A Social Anatomy of Fetal Surgery*, New Brunswick, NJ: Rutgers University Press.

Castro, F. (1967) *History Will Absolve Me*, Havana: Guairas.

Chomsky, A. (2000) 'The threat of a good example: health and revolution in Cuba', in J.Y. Kim (ed.) *Dying for Growth: Global Inequality and the Health of the Poor*, Monroe, ME: Common Courage Press.

Crabb, M.K. (2001) 'Socialism, health, and medicine in Cuba: a critical re-appraisal', doctoral dissertation, Emory University.

Davis-Floyd, R. and Sargent, C. (eds.) (1997) *Childbirth and Authoritative Knowledge: Cross-Cultural Perspectives*, Berkeley, CA: University of California Press.

Duden, B. (1993) *Disembodying Women: Perspectives on Pregnancy and the Unborn*, Cambridge, MA: Harvard University Press.

Feinsilver, J. (1993) *Healing the Masses: Cuban Health Politics at Home and Abroad*, Berkeley, CA: University of California Press.

Foucault, M. (1977) *Discipline and Punish*, trans. A. Sheridan, New York: Pantheon Books.

Franklin, S. (1997) *Embodied Progress: A Cultural Account of Assisted Conception*, New York: Psychology Press.

Gammeltoft, T. (2007a) 'Sonography and sociality: obstetrical ultrasound imaging in urban Vietnam', *Medical Anthropology Quarterly*, 21.2: 133-153.

— (2007b) 'Prenatal diagnosis in postwar Vietnam: power, subjectivity, and citizenship', *American Anthropologist*, 109.1: 153-163.

Georges, E. (2008) *Bodies of Knowledge: The Medicalization of Reproduction in Greece*, Nashville, TN: Vanderbilt University Press.

Greenhalgh, S. (1994) 'Controlling births and bodies in village China', *American Ethnologist*, 21.1: 3-30.

— (2003) 'Planned births, unplanned persons: "population" in the making of Chinese modernity', *American Ethnologist*, 30.2: 196-215.

Jordan, B. (1997) 'Authoritative knowledge and its construction', in R. Davis-Floyd and C. Sargent (eds.), *Childbirth and Authoritative Knowledge: Cross-Cultural Perspectives*, Berkeley, CA: University of California Press.

Kligman, G. (1998) *The Politics of Duplicity: Controlling Reproduction in Ceausescu's Romania*, Berkeley, CA: University of California Press.

Landsman, G. (1998) 'Reconstructing motherhood in the age of "perfect" babies: mothers of infants and toddlers with disabilities', *Signs* 24.1: 69–99.

Martin, E. (1987) *The Woman in the Body: A Cultural Analysis of Reproduction*, Boston, MA: Beacon Press.

Mitchell, L. M. and Georges, E. (1997) 'Cross-cultural cyborgs: Greek and Canadian women's discourses on fetal ultrasound', *Feminist Studies*, 23.2: 373-401.

Oaks, L. (2001) *Smoking and Pregnancy: The Politics of Fetal Protection*, New Brunswick, NJ: Rutgers University Press.

Rapp, R. (1999) *Testing Women, Testing the Fetus: The Social Impact of Amniocentesis in America*, New York: Psychology Press.

Rapp, R. and Ginsburg, F. (2001) 'Enabling disability: rewriting kinship, reimagining citizenship', *Public Culture*, 13.3: 533-556.

Rose, N. (2001) 'The politics of life itself', *Theory, Culture & Society*, 18.6: 1-30.

Rothman, B.K. (1989) *Recreating Motherhood: Ideology and Technology in a Patriarchal Society*, New York: Norton.

Taylor, J.S. (1998) Image of contradiction: obstetrical ultrasound in American culture', in S. Franklin and R. Ragoné (eds.), *Reproducing Reproduction: Kinship, Power, and Technological innovation*, Philadelphia, PA: University of Pennsylvania Press.

Urla, J. (1993) 'Cultural politics in an age of statistics: numbers, nations, and the making of Basque identity', *American Ethnologist*, 20.4: 818-843.

Van Hollen, C. (2003) *Birth on the Threshold: Childbirth and Modernity in South India*, Berkeley, CA: University of California Press.

Vidal, M. L, Padrón, G. and Gilpin, M. (1991) 'The development of high technology and its medical applications in Cuba', *Latin American Perspectives*, 18.2: 101-113.

Whiteford, L. and Branch, L. (2008) *Primary Health Care in Cuba: The Other Revolution*, New York: Rowman & Littlefield.

Endnotes

1   In 2010, Cuba reported its lowest-ever infant mortality at 4.5 deaths per 1,000 live births.

2   For a detailed examination of the effects of the embargo on Cuban health and medical practice, see the 1997 publication by the American Association of World Health (AAWH) based on a year-long study of Cuban medical institutions between 1995 and 1996.

3   I am deeply grateful to the doctors, particularly to Janet (a pseudonym), who facilitated my entrance into the clinic, shared their knowledge and experience, and made me feel so welcome.

4   The family doctors themselves estimated this number, as exact figures are impossible to obtain given the rapid changes currently taking place.

5   The few exceptions that I observed were in cases when the patients failed to appear for clinic visits. My research was conducted exclusively in Havana, and I cannot know for certain how faithfully these guidelines are followed in the rest of the island. However, given the political importance of maternal-infant health indices, it seems reasonable to believe that prenatal health care is a priority throughout the family doctor system.

6   Pregnancies are diagnosed through tactile exams at the family doctor clinic or through ultrasound, which is capable of imaging fetal sacs as early as four weeks of development. Such divergences from the techniques of pregnancy detection familiar in the U.S. highlight the peculiarities of Cuba's medical economy. While international prices and the U.S. embargo limit the availability of reagents for laboratory work, doctors' time and ultrasound machines are considered part of the "sunken costs" of the public health system.

7   By contrast with the high rate of infant mortality in developing countries from causes such as infectious disease and malnutrition, congenital malformations are the third leading cause of infant mortality in Cuba, after accidents and perinatal ailments.

8   I did not observe this personally since I was not able to observe in a genetic counseling consultation. While I do not doubt that some counseling may be more directive than acknowledged, high rates of abortion following positive genetic diagnoses should also be contextualized within Cuba's generally high abortion rate and low birth rate.

9   Despite efforts to promote sexual education and contraceptive use since the 1980s, Cuba's rate of abortions continues to be very high - a fact troubling to

the state and health professionals. Abortions are legal and easily available up to about 11 weeks gestation, and are performed by doctors in local polyclinics or hospitals.

9

Family Medicine, "*La Herencia*," and Breast Cancer: Understanding the (Dis)continuities of Predictive Genetics in Cuba[1,2]

Sahra Gibbon

Introduction

The multi-layered dynamics between genetic knowledge or technology and family relations are now being explored in a variety of ways across the social sciences. Early work examining how common sense ideas of heredity (Davison et al. 1989) and kinship (Richards 1996) inform notions of genetic inheritance have been extended to show the complex ways that family and kin relations are being "thrust into relief" (Featherstone et al. 2006) and, in some cases, themselves transformed by genomic interventions. A few of these studies point to the way that novel genetic knowledge is concretizing a "bio-genetic" conceptualization of the family that challenges diverse kin groupings, as well as the ideology of individual choice (Finkler 2000). Others highlight the way genomic knowledge can be linked to novel forms of what have been described as "biosociality" (Rabinow 1996) that give new significance or meaning to ideas of personhood and social relations in the family and the wider community (see for instance Silverman 2008).

Previous research undertaken in the UK, suggests that the particular arena of medicine characterized as "breast cancer genetics" offers an important context for examining the relationship between identity, genetic medicine and the family. Following the identification of the two inherited susceptibility genes, BRCA1 and BRCA2, in the mid 1990s the growth and emergence of this new area of clinical practice has provided a rich context for examining the relationships between genetics and kin or family relations, both inside and outside the clinic. This has revealed the often contradictory consequences of genetic knowledge for collective and individual identity. On the one hand, something of a productive fit between the field of medicine characterized as BRCA genetics and the family has been identified. This has been linked to gendered notions of female nurturance (Gibbon 2007a) or ideologies of the "traditional" family as constituted by ideas of bio-genetic relatedness (Finkler 2000). At the same time, other research has also revealed a tension between a particular individual's investment in what is perceived as preventative health and the relational consequences of predictive risk information for the family (Gibbon 2007b; Hallowell 1999). As the work of Konrad (2005) examining predictive genetic knowledge in the context of a different condition, Huntington's Disease, has demonstrated, genetic knowledge can play out unevenly in the family. Such studies illustrate how interdependent social and personal relations between kin can be negotiated as "rights to know" or "not to know" about predictive risk information (see also Featherstone et al. 2006).

Detailed case studies are now beginning to examine just how varied the consequences of genomic knowledge can be within the context of family or for community and social relations, depending on the disease condition and the cultural meanings that may be linked to it. While Kate Weiner (2011) notes how genetic risk information may have little consequence for biosocial relations for those with Familial Hypercholesterolemia, Margaret Lock (2008) points to the lack of novel biosocial organization around the genomics of Alzheimer's disease. She points out how this possibility is itself over-determined in part by the difficult task of family care-giving for an incurable and often devastatingly debilitating condition. As the work of Rayna Rapp (1999) demonstrates examining one very specific aspect of reproductive medicine - amniocentesis, there is still much to be done within Euro-American societies in understanding the way that genetic knowledges and technologies are informed by a variety of family contexts, class, gender, religion and ethnicity (see also Shaw 2000, 2009). Yet comparative anthropological studies *outside* of Euro-American contexts are also now demonstrating how very different notions of personhood and meanings associated with kin, family, and/or citizenship have specific consequences for the meaning of and degree of engagement with genetic information. More broadly, these studies point to the way differently situated moral values relating to spiritual, religious or community practices and institutional cultures, as well as state provision (or the lack) of health care may influence, facilitate or impede the application of novel biological knowledges and technologies (Gammeltoft 2007; Gibbon and Novas 2008; Gibbon and

Reynolds 2009; Inhorn 2008; Kampriani 2009; Sleeboom-Faulkner 2010).

This paper - drawing on critical comparative ethnographic research in the specific cultural context of Cuba - contributes to the task of illuminating the limits and varieties of "co-production" (Jasanoff 2004; Lock et al. 2000) in the relationship between genetic medicine and the family. With a public health system that has long aimed to provide comprehensive health care, based on the values of equitable and universal access, Cuba provides a unique arena for exploring the evolving relationship between genetic medicine and the family. Despite a lack of financial and technological resources to undertake wide-spread clinical predictive genetic testing for conditions such as breast cancer, there is an ongoing commitment in this context to mobilize what is described as "community genetics" as a public health endeavor. In Cuba this increasingly includes the collection and analysis of family history data linked to common complex diseases such as breast cancer. This paper examines practices of "community genetics" in Cuba, as this relates to an expanding interest in conditions such as breast cancer and the ongoing engagement in family medicine as part of state public health care provision. It demonstrates how Cuba provides an important comparative arena for exploring the continuities and differences in the relationship between genetic knowledge and technologies, personhood, kinship and the family.

Research Methods

The analysis in this paper is based on research that formed part of a collaborative project working with teams of Cuban genetic professionals in three different provinces in the east, center and west of the country at various time periods between 2006 and 2008. The most intensive period of research was undertaken between September 2007 and March 2008. All regions where the research was undertaken are areas with their own particular pre- and post-revolutionary history of socio-economic development; these are notably different to the Havana context (see Rosendahl 1997). Data collected included ethnographic findings working alongside Cuban community genetic health practitioners, visiting families in their homes as part of the routine collection of family history information and also use of a semi-structured questionnaire. The questionnaire was completed by 250 Cuban women in three different provinces of the country. Topics covered by the questionnaire, which included both open and closed questions, focused on health beliefs concerning the perceived causes of and risk factors for breast cancer, including genetic and what were described as "non-genetic" factors. The age range of participants completing the questionnaire was 16 to 80, and all were women. Of this group, half had in the past, or in a few cases were currently being treated for breast cancer. Sampling and recruitment of research participants was the responsibility of the Cuban collaborators. Liaising closely with local polyclinics and family doctors they identified and invited participants to take part in the research. No systematic selection of participants was made on the basis of a family history of breast cancer; however, approximately a quarter of research participants had family members affected by breast cancer or another type of cancer. Following the completion of informed consent procedures, the questionnaires were undertaken with participants mostly in their own home. Community genetic practitioners were present at all times and were responsible for writing in the research participants verbal response to the listed questions. The research was granted formal ethical approval by the PI's host institution, University College London in 2006. The data generated by the research was both qualitative and quantitative and is being analyzed in a variety of ways, using SPSS and ATLAS.ti data analysis software. Data presented in this paper draws from both the ethnographic component of the project and selected aspects of questionnaire data. This includes responses to a range of open-ended questions asking participants about their perceptions and beliefs concerning the causes of breast cancer, what participants perceived as the most important risk factors for the disease and their personal (if relevant) or general experience of breast cancer in the family or community. These responses have been analyzed based on a grounded theory approach drawing on a thematic analysis of re-occurring topics (Corbin and Strauss 1990) using ATLAS.ti software. While some other aspects of this research have been discussed elsewhere (Gibbon 2009; Gibbon et al. 2010), other findings will be published as analysis of the large and diverse data sets generated by the research are completed.

The first half of this paper describes the context and emergence of what is described as "community genetics" as part of a program of public health in Cuba. It suggests that there is significant continuity between the comprehensive provision of public health care and the recent expansion of community genetics that relates directly to the long-standing interest in and provision of what is described as "family medicine." The second half of the paper draws more directly on the questionnaire data with Cuban women, focusing on the cultural meaning of breast cancer and health beliefs associated with the causes of the disease. This includes a perception that breast cancer is caused by factors that arise from outside and impact on the body and the significance (or not) of genetic or what is understood as hereditary risk. This part of the paper reflects on the challenges to the practice of community genetics given both health beliefs relating to the causes of the disease and also the climate of fear and silence within the family associated with discussing a cancer diagnosis. The data presented in the second half of the paper suggests that despite the institutional culture that has long been focused on the health of the family, the meeting points between an expanding area of community genetics linked to identifying and acting on increased genetic risk of breast cancer, may not be so easily aligned and are in fact characterized by a range of discontinuities, tensions and differences.

Public Health, Family Doctors and Community Genetics

The development of a comprehensive public health system in Cuba emerged out of the success and commitments of the revolution in the late 1950s. Since that time, Cuban health care has long stood as an important symbol of the success of the Cuban socialist revolutionary efforts (Brotherton 2005; Feinsilver 1993). This process of transformation included the creation of local primary health care services in the 1960s and 70s with locally based polyclinics set up across the country, including areas where health care resources had previously been scarce or non-existent. In conjunction with this, tens of thousands of medical professionals were trained in the following decades, with official government health statistics suggesting in 2001 that there were 31,000 doctors or nurses or one for every 175 people (MINSAP Cuban Ministry of Public Health 2001). This enormous increase in medical professionals, combined with a focus on maternal health, has been seen as directly responsible for the high profile successes of the Cuban public health care system. This is particularly so with respect to reducing infant mortality and increasing life expectancy (Spiegel and Yassi 2004). As a result, despite the economic challenges of the ongoing U.S. embargo and the collapse of the Soviet subsidies in the 1990s, the epidemiological profile of the country has been transformed from one with "diseases of poverty to diseases of development," that now prominently include heart disease and cancer. According to the Cuban National Cancer Registry, over 2000 cases of breast cancer are diagnosed annually in Cuba with a population of about 11 million inhabitants, making the incidence about 40 per 100,000 inhabitants (Alvarez et al. 2003). Although there are regional variations within the country, breast cancer is the most common malignancy affecting Cuban women with incidence of the disease comparable to a global rate and increasing every year (Galán et al. 2009).

Many commentators have suggested that understanding not just the symbolic significance of the "revolution" in health care in Cuba over the last 50 years but also the logistics of its success, must be attributed to the system of so-called "family medicine" that began to emerge in the 1980s. The effort to provide more and better equipped family doctors who could attend to both the "physical and social well-being" of the Cuban population was consolidated by the state in 1984 as part of the Family Physician and Nurse Program (MEF) (Nayeri 1995). This laid out the plans for the current organization of primary health care with doctors working in local *consultorios* in the communities in which they lived. By 1995, this system was established over the whole country integrating hospitals, local polyclinic services and community-based doctors. This enabled Cuba to apply and, in the view of some, to a certain extent realize the principle of "health for all with a primary focus" (Spiegel and Yassi 2004: 97). Importantly, prevention and not just treatment was a vital part of this medicine in the community program, with health being seen somewhat holistically as a function of "biological, environmental and the social well-being of individuals" (Nayeri 1995: 324). This went far beyond clinical intervention to include "disease prevention, hygiene instruction, family planning and risk factor assessment" (Jenkins 2008: 13; see also Nayeri 1995; Brotherton 2005).

The emergence of "Community Genetics" in the last six years, as part of a national program of intervention, emerges directly out of the "holistic" and "preventative" focus on family medicine and a long standing, nearly 40-year old program of infant and maternal health. This is in part reflected in the way that many professionals now working in the field of community genetics previously worked as family doctors in their local communities, before retraining to become specialists in genetics. It is also reflected in the infrastructure for this new health focus and the way that it is being consolidated and linked to the system of primary care set up through programs such as the MEF, with networks of genetic centers and clinics. Employing a total of more than 1,600 persons across the country, many regional centers have their own dedicated genetic specialists, technicians and nurses (Teurel 2009). Such centers are often linked to polyclinics or local hospitals, but frequently also have their own designated buildings in residential areas with separate consulting rooms and laboratories.

The main day-to-day focus of work in these centers is newborn neonatal screening to monitor for rare chromosomal conditions and in helping to facilitate the national programs of prenatal screening for conditions such as sickle cell anemia. By comparison, the work of genetic teams in relation to complex adult onset conditions such as breast cancer is focused on the collection and collation of registries of families affected by such conditions. In some centers, the list of conditions for which family history was being collected included, at the time of the research, schizophrenia, Alzheimer's, heart disease, diabetes and more behavioral-type conditions such "alcohol addiction." With in total over 43,000 families forming part of a national registry, this is potentially a powerful resource for future genetic research and medicine (Teurel 2009). Yet due to the cost and lack of technological infrastructure, these centers are as yet unable to provide comprehensive clinical risk assessment or predictive information based on genetic testing for those with a family history of breast cancer. Newly established community genetic clinics are nevertheless engaged in the task of collecting family history data and identifying persons and families most at risk. It's important to note that such work was also propelled by the collaborative project that formed the basis of the research from which the data presented in this paper is derived. At the same time, this project was for Cuban collaborators essentially an "exploratory" study that would help assess and perhaps in the future expand the practices of community genetics in relation to breast cancer; it was also a means of enabling and facilitating the task of collating family history information.

Moving around rural and urban communities with different teams of medical geneticists in the three different provinces where the research for the project was undertaken, it was clear that the genetic professionals were very much at the center of their communities. Many had worked for years as family doctors living in or near the district in which they worked. Now as part of the program of Community Genetics, they could not walk down the street without encountering people they knew or more usually were stopped by people that recognized them. Frequent humorous comments were made by them about how their homes were like *consultorios* every night, with neighbors, friends and acquaintances calling by to ask advice about health problems. On one occasion walking back to the community genetic clinic in the residential area of a small town in an eastern province, the doctor I was with was recognized and stopped by a mother and her teenage daughter. This was to discuss the fact that the daughter might be pregnant, and would therefore need an abortion. She had stopped the doctor to ask whom she should see at the local polyclinic about this. With a mixture of dismay and dry humor about being so frequently stopped in the street and asked about such queries, Celeste the doctor laughingly said that was community genetics in practice – "*es genética comunitaria en realidad*." This social position at the heart of the communities where these health professionals lived was however particularly important in relation to the collation of family history details, as the experience of working with these health professionals illustrated.

It is true to say that the social context of health beliefs and practices in relation to breast cancer, the main focus of our collaboration, was of some interest to these health professionals. They were attuned to consider these aspects of health and well-being, primarily as a result of prior involvement in a broad based system of family medicine orientated towards a "holistic" preventative approach. Nevertheless, this involvement was for them also explicitly about the task of recording and registering family history or identifying high-risk families. It was not perhaps surprising therefore that they quite often literally took charge of these moments of completing the questionnaire, sometimes busily drawing up mini clinical family trees or firing further questions about dates or details relating to the history of cancer in the family. Yet unlike the sometimes tense atmosphere that such questions could generate in the clinical contexts in the UK, it soon became obvious that these were routine and expected questions for both practitioners and patients in Cuba. The ease with which such information was exchanged with medical practitioners was particularly evident when, as was frequently the case, other family members became part of the discussions in the hunt for details of the history of disease. This was illustrated one afternoon walking around a residential housing area with two members of the local genetic clinic, after completing the questionnaire with an elderly woman in a nearby block of flats. On encountering the nephew of the elderly research participant, whom the geneticist also knew, he was asked if he knew about a particular cousin's medical history, as his elderly aunt had been unable to remember. He was not surprised to be stopped in the street while cycling back home from work by the geneticist whom he also knew and responded with good humor and no hesitation to the request to clarify the details that his older relative had roughly sketched out.

Asking and giving information about family history is part of the routine patient/practitioner dynamic in Cuba, which the expanding field of "community genetics" taps into and builds from. As the previous example demonstrates, this was particularly evident in smaller communities where the relationship between genetic practitioners and the public was highly localized. Another illustration of this was the way genetic professionals would often become concerned that participants completing the questionnaires responded with what they perceived as the "correct answers." Similarly when questions elicited blank or non-responses from participants, the geneticists would comment with open consternation, sometimes attempting to prompt participants and stating as one practitioner said, "they do know the answers!"

Cuban geneticists, in their attention to the family, antenatal and newborn care, are like other doctors: powerful and embodied symbols of the revolution. The high profile program of "Internationalism" involving the export of thousands of doctors to Africa, Asia and South America for periods of one to three years has been central to this symbolic association of doctors with the revolution both within and outside Cuba (Feinsilver 1993). Importantly over the last few years, the first "medical missions" involving Cuban Community Genetic professionals to Venezuela have also taken place. This identification of genetic professionals with the ethics of the socialist state and "revolutionary values," linked to ideals of equality of access and universal care of the population, was reflected in the visual references in the genetic clinics themselves. Here handmade murals and public health messages would sit alongside José Martí poems and pictures of "Che," Fidel Castro and Hugo Chavez. As the collaborative project linked to the research was undertaken in different provinces, a number of health professionals revealed their commitment in working in this way. An event recounted from field notes illustrates how these sentiments manifested themselves for one community genetic practitioner.

Mayra's pride in showing me the newly built community genetics clinic, for which she is director, is evident. Unlike the generally old mainly very run-down buildings, the community genetics center stands out as a gleaming newly painted building on a hill overlooking a provincial town in one of the eastern provinces of the country. Its location next to the maternity hospital in part symbolically reflects the way that Cuban Community Genetics builds on and extends the success of a widespread program of maternal and infant health care. Inside the newly built center, the newness of the fixture and fittings is evident with work still going on to complete the center. It is also one of the few buildings in the town with air conditioning. As we are walking around the as yet unused freshly painted conference room, which will be used for regional meetings of community genetic health professionals, conversation drifts into discussion of how Mayra got to be director of the regional genetics clinic here in this small town. She talks of being a family doctor in the Sierra Maestra in the difficult so-called Special Period following the collapse of the Soviet subsidies in the late 1980s and how this inspired her in her work. Since then she always wanted to be able to come back to the town where she had grown up and work. After doing her training in genetics she welcomed the opportunity to set up the community genetics service in her home town. Mayra's dedication to the work is in fact noted by others who work in the center - one nurse making some humorous yet nevertheless pointed remarks had earlier said in a semi-ironic way how unlike the "other doctors" who have left Cuba, Mayra is a "good communist." Discussion with Mayra turns to talks of how she is going to put plants on the outside of the building to make the area more comfortable and welcoming for patients. She also mentions there is a sculpture that will be put up at the front of the building. She takes me to a small room where the sculpture that had been destined for this spot is currently being stored. It is a very classic representation of double-stranded DNA. But she tells me that they aren't going to use this one – she says it's "*feo*" – ugly. The sculpture that will now be placed at the front of the building is something much more abstract and organic rather than obviously an object representing DNA or scientific knowledge. The choice of public sculpture at the entrance to the newly built community genetic initiative seems to symbolically reflect an effort to represent and position the work of medical genetics as part of the larger, long-standing project of community public health care. That is as a normalized aspect Cuban health care that directly builds on a long-standing program of public health, rather than something obviously novel or different.

Long-standing investment and organization of public health in Cuba has in fact placed locally and community orientated family medicine at the heart of an endeavor, from which community genetics extends and builds from. This situation would also seem to provide a certain degree of leverage for the growth and expansion of genetic medicine in Cuba, including that linked to BRCA genetics. As the ethnographic material outlined above suggests, family doctors and community genetic professionals are a vital component of this endeavor, situated at the heart of locally organized system of health delivery that is centered on the family. At the same time, their commitment to integrating community genetics into the Cuban project of public health, while not always as uncritically supportive or unaware of the resource challenges to this endeavor in Cuba (see Gibbon 2009), is nonetheless central to the ongoing success and expansion of this field of medicine. The next section of this paper - drawing on the questionnaire data with Cuban women and focusing on findings relating to participants' health beliefs concerning inheritance and genetic risk - examines the diverse and somewhat uneven ways in which an institutional culture of family medicine and community genetics informs the meaning of breast cancer genetics in Cuba. The local institutional culture of Cuban health care outlined in the first part of this paper would seem to provide a fertile context for the continued expansion of community genetics as "family" medicine. Nevertheless, the findings from the questionnaire data related to health beliefs and the difficulties of talking about cancer in the family pose challenges to the implementation of predictive health interventions linked to breast cancer in the Cuban context.

Understanding the meaning and morality of breast cancer risk; family history, "la herencia" and "los golpes"

Analysis of aspects of the questionnaire data with Cuban women relating to open-ended questions concerning beliefs about the causes of breast cancer, risk factors for and the experience (if relevant) of the disease, points to the importance of a range of perceptions. In contrast to comparable research undertaken with clinical and non-clinical populations in the UK (see for instance Gibbon 2007a) the meaning of "breast cancer risk," as well as the morality normally associated with "health awareness" and engagement in preventative health interventions, appeared to be somewhat differently articulated (see also Gibbon 2009 and Gibbon et al. 2010 for further discussion of this contrast). Here two specific aspects of Cuban women's health beliefs about the causes of breast cancer are examined. First the way that genetic and hereditary factors were understood and discussed. Second the way that a physical "blow" or what was described as "*un golpe*" is seen as the primary cause of the disease. I argue that this is illustrative of the way that risk of developing the disease is mostly seen as arising from and impacting on the body, rather than being generated within the body or as the outcome of individual actions (or inactions). Elsewhere I've explored this finding in relation to the findings of different sets of data, such as the significance of dietary factors and the meaning of "stress" (Gibbon et al. 2010).

Given the relative unavailability of predictive genetic testing in Cuba, an absence of hype and hope-filled discussion of the BRCA genes in the media, as well as the virtual absence of a strong culture of breast cancer activism (certainly outside of Havana), it was perhaps not surprising to find that very few of the women completing the questionnaire had heard of the BRCA genes. Rather more surprising perhaps was the fact that many had no point of reference to the term "genes" or "genetic" factors. There would often be looks of bemusement and doubtful shaking of heads in response to open questions asking if people had heard of *los genes* or *la genética*. For those few persons who for whom the term "genes" was meaningful, discussion centered on a vague notion of something perhaps being transmitted in the blood. This was how a number of persons expressed this:

> *Piensa que es algo que se transmite de una familia a otra en la sangre.*
> (I think that it's something that is transmitted from one family to another in the blood.)

> *Los genes son algo en la sangre que dan herencia y entonces para cáncer también.*
> (The genes are something in the blood that is inherited and then the same for cancer also.)

It was significant that when similar questions about genetic risk were rephrased in terms of "hereditary factors" (*factores hereditarios*) there was a much more widespread positive recognition. That is to say there was much more likely to be discussion and understanding that *"la herencia"* (inheritance) and *"la salud o las enfermedades"* (the health or illness) of *"los antecedentes"* or *"antepasados"* (ancestors) might contribute to the risk of disease. This suggested that there was a particular cultural salience in this context surrounding hereditary risk, if not genes, genetic factors or more specifically the BRCA genes.

This was particularly evident in the way that direct questions relating to genes, as opposed to hereditary risk factors, could elicit both strongly negative and positive responses often from the *same* person. This was subtly illustrated in this respondent's comments. Talking about where she had heard about the link between hereditary factors and breast cancer, she said:

> *[E]n documentales por la televisión, en conversaciones con personas se habla de que es un factor importante porque hay varias personas en las familias afectadas. De como es que los genes producen cáncer no lo he escuchado.*

> (]I]n television documentaries on the television, in conversations with people I've heard that its an important factor when there are different family members affected, but I haven't heard anything about how genes produce cancer.)

Another woman was more cautious in her response:

> *[P]udiera ser una causa genética, no ha escuchado en específico sobre los genes en respecto de cáncer de mama.*

([Y]ou could say it was genetic but I haven't heard specifically about genes for breast cancer.)

Other persons were more definitive, with some incredulous in response to the suggestion that breast cancer was linked to hereditary or genetic factors, "*la herencia no está vinculado con cáncer de mamá*" ("inheritance isn't linked to breast cancer"). In another instance two sisters who had both in fact had breast cancer talked much about their shared experiences of living through the treatment, "*somos gemelos en respecto de cáncer de mamá*" said one of them ("we are 'twins' when it comes to breast cancer"). However, they both refused and strongly refuted when prompted that there was anything hereditary in the fact that they had both had breast cancer.

One woman quite explicitly in response to a query about genes linked to breast cancer made a clear distinction between unknown genetic factors and known hereditary risk:

No tengo conocimiento de esto [factores genéticas]pero piensa que pueda ser hereditario o sea que uno nazca con eso y se manifeste a cualquier edad.
(I don't know anything about this [genetic factors] but I think that it could be hereditary or it's that you are born with it and it can appear at any age.)

While BRCA genes and genetic factors more generally had little point of reference for many research participants, hereditary risk factors were, as these examples suggest, more readily associated with the increased incidence of the breast cancer. That is, while some felt that factors such as having a family history might be important in the development of breast cancer specifically, there was nothing like the kind of reading of breast cancer as a "genetic" disease that accompanied the high-profile announcements that heralded the hyped and hope-filled discovery and application of new knowledge of the BRCA genes in the mid and late 1990s in the UK and other Euro-American societies (Gibbon 2007a; Pathasarathy 2007). It was also not insignificant that in responding to the questions, conditions such as diabetes and asthma were in fact much more readily and easily linked to hereditary factors than breast cancer; both are diseases that Cuban public health care directly attends to through the program of family medicine. In summary, the disjuncture between what were perceived as more meaningful hereditary factors and what might be described as unknown or unknowable genetic factors suggested that while the former had a particular meaningful resonance in peoples' lives, the latter did not.

This situation must in many ways be read in relation to the longstanding institutional culture and ideological values imparted through the Cuban project of public health care and the system of family medicine that has been in place in Cuba for the last 20 years. At the very least, this ensures that attending to the history of disease in the family is commonplace. As one questionnaire respondent said in response to a query about why she believed hereditary factors were important in relation to breast cancer, "*porque en las consultas siempre le pregunten si tienen antecedentes familiares con enfermedades*" ("because in consultations they're [doctors] always asking if you have relatives who are sick"). This was more succinctly and directly interpreted by one genetic professional who said, "people are so used to questions about family history or thinking that it is important, because we are always asking them about it."

It was certainly notable that in responding to direct queries about family history from genetic professionals many persons willingly exchanged such information. Moreover, many had an impressive grasp of the details of their family medical history, remembering not only the dates a relative had died or been diagnosed but sometimes the details of the medical procedure they or a relative had undergone. Elsewhere I've argued that this impressive ability to be conversant in and recount the family medical history or be engaged with biomedical procedures, histories and scenarios, which often constituted a response to queries about the *experience* of breast cancer, reveals the extent to which *biologized* citizenship may be at stake in the Cuban arena (see Gibbon 2009).

Nevertheless, examining in more detail the particular way in which "embodied risk" linked to breast cancer is understood by individual Cuban women provides another context for understanding the challenges and discontinuities in the translation of predictive health interventions linked to breast cancer in the Cuban context.

It is true to say that there was some discussion by a few research participants regarding the need to take care of one's own personal health and that to neglect this was perceived as being detrimental to well-being and might lead to disease. This was particularly evident for those few participants who lived or worked in or near tourist regions or who, because of family living abroad, had access to some limited but nevertheless welcome extra financial resources. Nevertheless, a moral discourse about *individual* responsibility for health, which previous research in the UK had suggested was central to interest in and patient mobilization around BRCA genetics (see Gibbon 2007a), was a far from obvious terrain of discussion for the majority of those who took part in the research.

For example, so-called "lifestyle risk factors," although acknowledged by some as important to overall health, were not always seen as factors that individuals could easily personally or directly alter or affect. In general there was a feeling that the strongest risk or danger came from *outside* of and impacted *upon* the body. As I've argued elsewhere, this often meant identifying risk factors which were not only outside of the control of the individual but sometimes outside of the control of the Cuban state. This might include pollution from international conflicts in Iraq or Afghanistan and environmental contaminants or ozone depletion (Gibbon 2009) or a "deficit" in dietary food intake (Gibbon et al. 2010*)*. Here I explore this rendering of embodiment in relation to perception of risk related to breast cancer by examining the frequency and manner in which respondees explained the cause of the disease in terms of a "blow" or in Spanish *un golpe*.

It was notable that in response to questions about the causes or risk factors, more than half of the total number of respondents thought that a *golpe* was the primary or a secondary cause of or a factor in the development of cancer. While such descriptions were sometimes used interchangeably to reference both a physical blow that might have caused the disease or psychological trauma (sometimes also used to describe the experience of having breast cancer) it was the former meaning that was most evident in the response of participants.

For some it was a simply that a "blow" was sufficient enough to "trigger" a cancer, *"un golpe puede desencadenar el cáncer de mama."* For others a blow might be related to hereditary factors in more explicit ways. For instance this could be used to emphasize the causative function of the former, *"no lo relaciona con la herencia, lo relaciona con el golpe que recibió"* ("it's not related to inheritance, its related to a blow that was received"). At other times, this reasoning was inverted but in ways which still served to emphasize the importance of a *golpe* in understanding the cause of breast cancer:

> [P]*iensa que los mas importante es el factor hereditario, porque no todos en su familia han tenido golpes*
> ([S]he thinks that its related to hereditary factors, because not everyone in the family has had a blow [to the breast].

For some it was the fact that a physical blow might not be attended to in time which was perceived to be the problem. As one woman put it, *"los golpes que no se atienden bien pueden originar un coagulo y de ahí un cáncer"* (the blows that aren't looked after can develop a clot and then from there a cancer). This kind of statement reflected to some degree the importance of taking care of oneself, or ensuring that medical attention was sought. It also suggests that at least for some, individual health awareness was not totally absent to this kind of reasoning.

It is perhaps notable that particular ideas of female gender were sometimes caught up with discussion of the way a "blow" to the breast was perceived as a risk factor for breast cancer. There was frequent mention that the breast was a *"zona delicada para las mujeres"* – a delicate and sensitive area of the body for women which was susceptible to injury. More telling was the way inappropriate activity for women or what was described as *"fuerza física"* or "physical force" was also implicated in the development of cancer. In the opinion of some, this could be linked to the novel need for women to undertake physical work as a result of economic demands following the Special Period, or it might refer to women's recent involvement in traditionally male sporting activities, such as boxing or weight building. Such perceptions would seem to reflect, in part, anxieties about the changing role of women in Cuban society. Highly illustrative of such feelings was a comment from one participant who believed it had been the blows she had received from a rifle during the pre-revolutionary struggle she had participated in during the late 1950s that had caused her cancer, *"el mas importante creo que es un golpe, yo recibí golpes cuando la clandestinidad me golpearon con la culata de un fusil."* ("I think the most important is a blow I received when a I was hit by rifle butt by a secret campaigner.")

Others working in the United States and Mexico have noted that *"un golpe"* is not an uncommon explanation for breast cancer, particularly among Hispanic (and also some non-Hispanic) populations (Finkler 1991). Hunt points out in her work in Mexico that a *golpe* may perhaps function like the notion of "stress" in the west, providing a "conceptual bridge" and form of "moral reasoning" between disorder in the body and society and therefore, like the notion of stress, provide "a flexible, versatile symbol, locating the source of the disorder within an individual life history." But as Hunt also acknowledges, it importantly locates health risk in terms of "an attack *from without"* (my emphasis Hunt 1998: 304). This suggests something slightly different from certain "western" readings of stress that would normally center on the culpability of the individual.

The frequency of recourse to explanations about the causes of breast cancer or risk of developing the disease as being linked to "a blow" provides one illustration of the way that for many Cuban women the most important etiological disease causing agents are those that impact *on* the person, body or self, rather than being generated from within or which arise as a result of an individual's own actions. Such readings seem to reflect a perception of cancer as not a product of "the body at war with itself," as is common in western readings of cancer, (Sontag 1991; Stacey 1997) but as being under attack or as result of the actions of "impersonal" outside agents. This might partly be understood in relation to the context of economic shortage in Cuba, which commenced with what is commonly referred to as the Special Period following the collapse of the Soviet subsidies in the 1990s. Some suggest that this climate of shortage has helped to foster a culture of capitalism in necessary practices of barter and exchange (Brotherton 2005, 2008). The qualitative data presented here relating to health beliefs about breast cancer among participants outside of the metropole of Havana suggests this climate also informs perceptions of embodied "future" risk in ways that do not necessarily provide a viable context for an expanding field of predictive genetics.

A recent prominent discourse within social science, which has emerged partly in response to and as a way of understanding the meaning and significance of identity in the context of genomics, has suggested and implied that there is strong ideological fit between the emergence of genetic knowledge and the expansion of what has been described as "self-actualizing" personhood. In other words the idea that there is a moral obligation to take responsibility for one's health is something of a pre-requisite for, as well as a consequence of, novel biomedical technologies and genomic medicine (see for instance Rose and Novas 2005). That is, a particular form of "biological citizenship" linked to a burgeoning culture of breast cancer activism which mobilizes a preventative health ideology in its emphasis on individual vigilance and awareness, seems to have been an important aspect of the growth and expansion of breast cancer genetics in the UK and U.S. (Gibbon 2007a, 2007b; Parathasarthy 2007). There is also evidence that an emphasis on female health awareness is a more complex but nevertheless important feature of the way this field of medicine has expanded elsewhere, as Kampriani (2009) demonstrates in her work on Greece. In Cuba, such public or individualized health activism, with respect to breast cancer, seems somewhat absent and differently configured. That is, for the most part it is not only that the actions of individuals seem not to be perceived as the primary cause of diseases such as cancer, but that the most dangerous agents are located outside the body. This does not mean that a notion of individual responsibility is totally absent in Cuban public health discourse, which is itself dynamically responding to a changing political context and climate of health provision (see Brotherton 2005). There were instances, as mentioned previously, in undertaking the questionnaires in certain parts of the country closer to the metropole of Havana or tourist regions in the country where evidence of a discourse of health awareness informed research participants responses to questions about risk (see Gibbon et al. 2010). Despite an awareness of and interest in "hereditary factors," the critical differences in the way that embodied risk is configured by Cuban women who took part in the questionnaire study does suggest that this is a context in which one of the apparent requirements for the expansion of predictive genetic medicine are less readily and immediately visible.

The final section of this paper further illuminates this aspect by pointing to one of the difficulties that confronted genetic practitioners as they moved in their local communities collecting family history information about feared and still stigmatized diseases such as cancer.

From community genetics to predictive medicine. The challenge of talking about "cancer" in the family

While the long standing institutional culture of family medicine would seem to provide a logical starting point for the expansion of community genetics in Cuba to encompass predictive interventions, the reading of "breast cancer risk" by Cuban research participants would suggest that this transition is not so easily achieved. Ethnographic research working with health professionals highlighted an issue that became particularly evident in the course of undertaking the questionnaires with families in their homes; that is the challenge of openly discussing the diagnosis of "cancer" in the family. Given the easy exchange of family medical history information that the practice and culture of community genetics and family medicine in Cuba would seem to facilitate, this difficulty seems somewhat contradictory. It reflects the dread and fear in Cuba that is associated with the modern disease of cancer and an ongoing culture of paternalism that affects public health practices.

The ability of health practitioners to access and willingness on the part of patients and their families to provide information about family health history and other diseases is an essential aspect of the development of predictive genetic interventions. Even in Euro-American contexts, where the ability to carry out genetic testing is more widely available, accessing family history information is still of prime importance in risk assessment procedures, in decisions about whether to offer testing and in helping to establish the meaning of mutation and predictive testing (Guttmacher et al. 2004). The absolute need for collective family engagement in predictive genetic medicine is perhaps most readily evident when the requirement to share information and pass on risk information breaks down in the family, impeding and sometimes preventing the pursuit of a genetic risk diagnosis for different members of the family (Gibbon 2007b; Hallowell 1999; Konrad 2005). Despite the fact that the institutional culture of family medicine in Cuba seems to facilitate the collection of family history information and potentially at least the future expansion of this arena of medical intervention, the inability often to discuss and share a diagnosis of cancer between a patient and their family poses a significant practical and ethical challenge for community genetics in Cuba. That is, while a family history of medical procedures and hospital interventions may have been widely known and remembered by patients and their families, the specific diagnosis of cancer was sometimes couched in more metaphorical terms or more problematically simply absent from individual and collective conversations.

This was illustrated on a number of occasions when working with Cuban health professionals. On such occasions on arriving at the home of a potential research participant, we would be met by an anxious relative of the chosen research participant. While they were often happy for the research questionnaire to be given to the person in question, they were concerned about their relative (normally in these situations a mother, grandmother or aunt) being told that they currently had or had in the past been diagnosed with cancer or "*el cangrejo*" – the crab, as sometimes the disease was metaphorically described. On some occasions this difficulty led to a decision to not undertake a questionnaire. In other moments when it was less clear if a relative had been told or not if they had or had had cancer, open-ended questions were re-phrased by community genetic practitioners to avoid the use of the word 'cancer'. In these situations, "*problemas con las mamas*" (breast problems), "*nódulos*" (breast lumps) and other euphemistic terminology would be used in circuitous ways by practitioners, relatives and sometimes also research participants themselves in responding to questions. As recent studies in Cuba have suggested, the tradition of non-disclosure of cancer diagnosis to the patient (Roll et al. 2009) reflects a degree of long-standing paternalism in the public health system. This, coupled with a dynamic of care within the family that assumes that not telling a relative their diagnosis of cancer is the best course of action, constitutes a series of problematic challenges for genetic practitioners in undertaking many aspects of their work. This includes the requirements for meeting ethical guidelines relating to individual informed consent (see Gibbon 2009). An inability to provide widespread predictive testing ensures that in part the problems of sharing future genetic risk information in the family do not constitute a significant challenge as yet in the Cuba. Nevertheless, the problems of discussing a diagnosis of cancer within and between the patient and their family reflect a significant point of rupture and discontinuity in the translation of predictive medicine in the Cuban arena.

Conclusion

This paper, drawing on fieldwork with a cohort of Cuban women and working with Cuban genetic health professionals, has explored some of the uneven and disjunctured dynamics that characterize the translation of genetic medicine linked to breast cancer in a very specific national/cultural arena. It has been argued that the continuities and discontinuities that characterize this process must be situated in relation to a nexus of social practices and cultural discourses that both enable and challenge this endeavor in different ways. This includes the long-standing institutional culture of Cuban public health care provision, research participants' beliefs or perceptions about risk relating to breast cancer, and the challenge of openly discussing a cancer diagnosis between patients or health practitioners in the context of the family.

In one sense, the Cuban case seems to point to the presence of particular continuities in the practice and provision of public health and novel interventions related to assessing genetic risk in common complex conditions such as breast cancer. Ethnographic evidence suggests that long-standing locally organized aspects of the health care system, focused on family medicine, have helped to give meaning and significance to the idea of family history as a risk factor for disease whilst also furthering the emerging medical arena of community genetics. That is, the work of collating family history and managing the delicate inter-familial relations that are part of genetic interventions is central to a tacit practice of care-giving within the community. Here doctors are situated as paternal guardians and gatekeepers of this and other health information. Although without technological or financial resources to undertake widespread predictive testing, being able to match genealogical data to clinical records confers a degree of scope and flexibility to an emerging and changing practice of Cuban medical genetics.

The second half of this paper highlights how alignments between the focus on family medicine and moves to incorporate predictive health interventions in Cuba are in fact overlaid by difficulties and tensions that act as impediments to the easy translation of BRCA genetics in this context. Three such aspects of these dynamics have been explored in this paper. Drawing on qualitative data examining health beliefs in relation to breast cancer, the findings presented here suggest that while family history may be perceived to constitute a risk factor for many diseases (not just breast cancer), "genetic risk" has little meaningful resonance for many research participants. At the same time, particular perceptions of risk are informed less by a discourse of individual moral responsibility than by an understanding of risk and danger that is more likely to be located in cancer causing agents that exist *outside* or operate *on* the individual. This paper has drawn on illustrative examples relating to the perceived causative effect of physical blows to the breast and gendered ideas about excessive and what is understood as "unnatural" physical activity undertaken by women. While further research and analysis of the large and diverse data set will further illuminate this finding, the data presented here highlight the need to examine the varieties of biosocial identities that are at stake in the translation of genomic technologies across a diverse national and transnational global terrain (Gibbon et al. 2010). The final section of the paper points to a further difficulty linked to the effort to incorporate predictive medicine, which somewhat conflicts and would appear to be at odds with the institutional culture of family medicine in Cuba. Here the silences and use of metaphorical language that can characterize medical and family discourse relating to a diagnosis of cancer constitute a significant ethical and logistical challenge to the translation of predictive interventions linked to increased risk of disease such as breast cancer.

In summary, the very different configuration of factors that provide a context for the emergence of breast cancer genetics in Cuba provides a powerful illustration of the need for broader comparative perspectives in examining the relationship between genetic interventions and the family. The data presented here demonstrate the importance of examining the way that culturally and historically specific variables must be accounted for, as the work of translating predictive medicine is undertaken across comparatively different national arenas. This includes understanding the continuities, disjunctures and differences at stake in the dynamic relationship between genetic medicine, identity and the family.

References

Alvarez, Y.G., Garrote, L.F., Torres Babie, P., Guerra, M. and Jordan, M.G. (2003) 'Breast cancer risk in Cuba', *Meddic Review*, 5:2-3.

Brotherton, S. (2005) 'Macroeconomic change and the biopolitics of health', *Journal of Latin American Anthropology*, 10.2: 339-369.

Brotherton, S. (2008) '"We have to think like capitalists but continue being socialists": medicalized subjectivities, emergent capital and socialist entrepreneurs in post-Soviet Cuba', *American Ethnologist*, 35.2: 259-274.

Corbin, J., and Strauss, A. (1990) 'Grounded theory research: procedures, canons, and evaluative criteria', *Qualitative Sociology*, 13:3-21.

Davison, Charlie, Frankel, S. and Davey-Smith, G. (1989) 'Inheriting heart trouble: the relevance of common-sense ideas to preventive measures', *Health Education Research Theory & Practice*, 4.3: 329-340.

Featherstone, Katie, Atkinson, P., Bharadwaj, A and Clarke, A. (2006) (eds.) *Risky Relations: Family and Kinship in the Era of New Genetics*, London: Berg.

Feinsilver, J. (1993) *Healing the Masses: Cuban Politics at Home and Abroad*, Berkeley, CA: University of California Press.

Finkler, K. (1991) *Physicians at Work, Patients in Pain. Biomedical Practice and Patient Response*, Boulder, CO: Westview Press.

— *Experiencing the New Genetics:Family and Kinship on the Medical Frontier*, Philadelphia, PA: University of Pennsylvania Press.

Galán, Y., Fernández, L., Torres, P., and García, M. (2009) 'Trends in Cuba's cancer incidence (1990 to 2003) and mortality (1990 to 2007)', *Meddic Review*, 11.3: 19-26.

Gammeltoft, T. (2007) 'Prenatal diagnosis in postwar Vietnam: power, subjectivity, and citizenship', *American Anthropologist*, 109: 153-163.

Gibbon, S. (2007a) *Breast Cancer Genes and the Gendering of Knowledge: Science and Citizenship in the Cultural Context of the 'New' Genetics*, London, Basingstoke: Palgrave Macmillan.

— (2007b) 'Genealogical hybridities: the making and unmaking of blood relatives and predictive knowledge in breast cancer genetics', in J. Edwards, P. Wade and P. Harvey (eds.) *Anthropology and Science: Epistemologies in Practice*, Oxford: Berg.

— (2009) 'Genomics as public health? Community genetics and the challenge of personalised medicine in Cuba', *Anthropology and Medicine Special Issue: Biomedical Technology and Health Inequities in the Global North and South*, 16.2: 131-147.

Gibbon, S., Kampriani, E., zur Nieden, A. (2010) 'BRCA Patients in Cuba, Greece and Germany: comparative perspectives on public health, the state and the partial reproduction of 'neo-liberal' subjects', *Journal of BioSocieties,* 5: 440-446.

Gibbon, S. and Novas, C. (eds.) (2008) *Biosocialities, Genetics and the Social Sciences: Making Biologies and Identities,* London: Routledge.

Gibbon, S. and Reynolds, S. (eds.) (2009) 'Introduction', *Anthropology and Medicine Special Issue: Biomedical Technology and Health Inequities in the Global North and South,* 16.2: 131-147.

Hallowell, N. (1999) 'Doing the right thing: genetic risk and responsibility', in J. Gabe and P. Conrad (eds.) *Sociology of Health & Illness Monograph 5: Sociological Perspectives on the New Genetics. Sociology of Health & Illness,* 21: 597-621.

Hunt, L. (1998) 'Moral Reasoning and the Meaning of Cancer: Causal Explanations of Oncologists and Patients in Southern Mexico', *Medical Anthropology Quarterly,* 12.3: 298-318.

Inhorn, M. (ed.) (2008) *Reproductive Disruptions: Gender, Technology and Biopolitics in the New Millenium,* Oxford, New York: Berghahn Books.

Jasanoff, S. (2004) *States of Knowledge: the Co-Production of Science and the Social Order,* London: Routledge.

Jenkins, T. (2008) 'Patients, practitioners and paradoxes: responses to the Cuban health crisis of the 1990s', *Qualitative Health Research,* 18.10:1384-1400.

Kampriani, E. (2009) 'Between religious philanthropy and individualised medicine: situating inherited breast cancer risk in Greece', *Anthropology and Medicine Special Issue: Biomedical Technology and Health Inequities in the Global North and South,* 16.2: 147-165.

Konrad, M. (2005) *Narrating the New Predictive Genetics,* Cambridge: Cambridge University Press.

Lock, M. (2008) 'Biosociality and susceptibility genes: a cautionary tale' in S. Gibbon and C. Novas (eds.) *Biosocialities, Genetics and the Social Sciences: Making Biologies and Identity,* London: Routledge.

Lock, M., Young, A., and Cambrosio, A. (eds.) (2000) *Living and Working with the New Medical Technologies: Intersections of Inquiry,* Cambridge: Cambridge University Press.

Ministerio de Salud Publica (MINSAP) Republica de Cuba (2001) *Annuario Estadístico.* La Habana, Cuba.

Nayeri, K (1995) 'The Cuban health care system and factors currently undermining it', *Journal of Community Health*, 20.4: 321-335.

Parathasarathy, S. (2007) *Building Genetic Medicine: Breast Cancer, Technology and the Comparative Politics of Health Care*, Cambridge, MA: MIT Press.

Rabinow, P. (1996) 'Artificiality and enlightenment: from sociobiology to biosociality', in J. Crary and S. Kwinter (eds.) *Zone 6: Incorporations*, New York: Zone.

Rapp, R. (1999) *Testing Women, Testing the Fetus: The Social Impact of Amniocentesis in America*, New York: Routledge.

Richards, M. (1996) 'Lay and professional knowledge of genetics and inheritance', *Public Understanding of Science*, 5: 217–230.

Roll, I.V., Simms, V. and Harding, R. (2009) 'Multidimensional problems among advanced cancer patients in Cuba: awareness of diagnosis is associated with better patient status', *Journal of Pain and Symptom Management*, 37.3: 325-330.

Rosendahl, M. (1997) *Inside the Revolution: Everyday Life in Socialist Cuba*, Ithaca, NY: Cornell Univeristy Press.

Rose, N. and Novas, C. (2005) 'Biological citizenship' in A. Ong and S. Collier (eds.) *Global Assemblages: Technology, Politics and Ethics as Anthropological Problems*, Oxford: Blackwell.

Shaw, A. (2000) 'Conflicting models of risk: clinical genetics and British Pakistanis', in P. Caplan (ed.) *Risk Revisted*, London and Sterling, VA: Pluto Press.

— (2009) *Negotiating Genetic Risk: British Pakistani Experiences of Genetics*, Oxford and New York: Berghahn Books.

Silverman, C. (2008) 'Brains, pedigrees and promises: lessons from the politics of autism genetics', in S. Gibbon and C. Novas (eds.) **Biosocialities, Genetics and the Social Sciences**: *Making Biologies and Identity*, London: Routledge.

Sleebom-Faulkner, M. (2010) *The Frameworks of Choice: Predictive and Genetic Testing in Asia*, Amsterdam: University of Amsterdam Press.

Sontag, S. (1991) *Illness as Metaphor: And, AIDS and its Metaphors*, London: Penguin.

Spiegel, J. and Yassi, A. (2004) 'Lessons from the margins of globalization: appreciating the Cuban health paradox', *Journal of Public Health Policy*, 25.1: 85-110.

Stacey, J. (1997) *Teratologies: a Cultural Study of Cancer*, London: Routledge.

Teruel, B.M. (2009) 'Cuba's national medical genetics program', *MEDICC Review*, 11.1: 11-13.

Weiner, K. (2011) 'Exploring genetic responsibility for self, family, kin in the case of hereditary raised cholesterol', *Social Science & Medicine*, 72.11: 1760-1767.

Endnotes

1   This article was originally published as Gibbon, S. (2011) 'Family medicine, "La Herencia" and breast cancer; understanding the (dis)continuities of predictive genetics in Cuba', *Social Science & Medicine*, 72: 1784-1792, and is reprinted with permission.

2   The research presented here was funded by a Wellcome Trust post-doctoral fellowship award number WT068432M. My sincere thanks to all the Cuban participants who took part in this research and to those who made it possible to complete this research. I am also grateful to the anonymous reviewers who provided invaluable assistance in commenting on the developing drafts of this paper.

10

Masculinity and Sexuality in Cuba: Myths and Realities[*]

Julio César González Pagés

A Distinguished Member

One of the most hotly debated and controversial topics in the field of masculinity studies is men's relationship to their sexuality. We have an extensive mythology associated with the sexuality and perceived behavior of Cuban men, embellished with imaginaries of exceptional penile attributes.

The relationship between man and his penis goes beyond sexual or biological issues. The Latin American culture of masculinity worships the penis excessively, using many euphemisms for the organ, nearly all of which evoke powerful, firm objects.[1] This is the expectation that boys must have from the time they discover their "member," one of the most common euphemisms for the penis which also leaves no doubt about its position in the hierarchy and the excitement it generates once it becomes visible.

A study conducted by the Cuban journalist Aloyma Ravelo (2005), entitled *Sexo Tropical: El tamaño del pene en la imaginería de estudiantes universitarios de La Habana* (Tropical Sex: Imagined penis size among Havana university students), states: "From boyhood Cuban men are socialized to prove their manliness and sexual power based on the size of their penis" (240).

Having a large penis opens the door to the future adult male's sexuality because greater diameter and length are associated with higher virility. In three surveys conducted in masculinity workshops in the City of Havana, many of these criteria were confirmed. The survey population consisted of 173 male and 57 women from various professions with a high school or college education and the following racial and age demographics: 119 White, 88 Black, 23

[*] Translated by Rocky Schnaath

Mulatto or Asian; age 22 to 45. In terms of the penis size myth, 71% of those surveyed rated Black men as having the largest penises, which they attributed to their genetic strength and African ancestry.

The questionnaire revealed that 65% of the women surveyed preferred men with large penises, the opposite of what we heard from women when we spoke with them individually in many conversations prior to the survey, when they said penis size did not matter, whereas spiritual values did. This kind of contradiction illustrates the complexity of contemporary cultural imaginaries, as well as the need to approach them from a holistic point of view.

This contradiction has been written about in Western literature since the libertine movement of the 18th century, championed by the Marquis de Sade in France. In "Philosophy in the Bedroom," he advises from the opening line: "the mother will prescribe its reading to her daughter"; later he directs a speech to men and women:

> Voluptuaries of all ages and sexes, it is to you only that I offer this work; nourish yourselves upon its principles: they favor your passions, and these passions, whereof coldly insipid moralists put you in fear, are naught but the means Nature employs to bring man to the ends she prescribes to him; harken only to these delicious promptings, for no voice save that of the passions can lead you to happiness. (Sade 1990: 8)

The phallocentric approach used in this work by de Sade represents a masculine design based on biological differences in which the world revolves around the penis. Researcher Victor Seidler (2000) has questioned the Enlightenment for proposing the identification of masculinity with reason and the organization of society around men's interests.

The lack of relationship of penises to the current cultural aesthetic does not allow the male body to be integrated into the arts without setting aside the morbid interest or sadomasochistic iconography suggested by U.S. artist Robert Mapplethorpe. (Morrisroe 1996)

When I was in Barcelona in June of 2003, I went to see a Spanish performance of the play *Puppetry of the Penis* by the Australians Simon Morley and David Friend. The play deals with the abilities of the penis, which the actors turn into actual marionettes in the form of the Eiffel Tower, the Loch Ness monster, a pelican, and a hamburger. Once more the penis was presented as an icon of power. The only woman on stage, actress Roser Pujol, told the Spanish newspaper *El País*: "*Puppetry of the Penis* is highly recommended to teach the average housewife to see sex as a much more natural thing; it has thera-peutic value" (Ginart 2003: 34). The actress' opinion once again leaves women reflecting on how to please the penis; the men, however, remain ignorant of their bodies, their sex organs and their relationship to sexual enjoyment with their partner.

However, many men who disagree with this kind of performance dare to react as Vicente Verdú (2003) did in his commentary "The Penis and its Shadow," published in the Spanish newspaper *El País*, in which he states, "For those feminists who still insist on complete equality, here is the difference. If there were a play in which the female body were treated the same way the male body is in *Puppetry of the Penis*, even Miriam Tey would roll over in her grave" (38). In his review, Verdú takes on feminists, who are paying the price for men's lack of creativity with their bodies in general and their penises in particular.

In Cuba, without the same intention of penis worship, the play *La Celestina*, performed by the theater group El Público, brought male nudity to the stage. The play drew an unusual number of young people, and others who do not usually attend this kind of performance confessed that they "had gone to see it because some television actors were appearing nude."[2]

Far from helping to demystify these imaginaries, such trends in the cultur-al industry exacerbate them. This is confirmed by the specialist Demian Ruiz (2001) in his commentary "Oddities of the Penis" published in the magazine *Men's Health*: "You probably look at your penis in the same way a spinster looks at her cat: you think its particular attributes make it unique and extraor-dinary" (90-93).

Such insights abound in magazines targeting men, which are sold to teach us to care for and love ourselves. In this sense, it is truly remarkable that these values extend to all cultural manifestations, accompanied by the same *machista* biases, all over the world.

Another manifestation of hegemonic masculinity is public masturbation, which is one of the options men choose in order to give free rein to their erotic and morbid dreams or to express frustration about their inability to ex-press their sexuality, almost always violating women's space and forcing them to leave in the presence of male aggression. Havana, like many cities around the world, has nocturnal venues no woman would dream of frequenting for fear of such acts of aggression. Speaking with a group of university students about why the opposite was not true – why we do not see women on the beach, in movie theaters and on dark streets exposing their genitals and mas-turbating so men would see them – they laughed and said that was crazy and would never happen. The pure irony of this does not escape me.

Male masturbation in Cuba is part of a rite of initiation, and when the time comes, jokes like, "Let it go!" or "I'm going to put a bell on your hand" can be heard when knocking on the bathroom or bedroom door.[3] This is never a sign of non-acceptance, and it indicates that something very important, which is an affirmation of his masculinity, is occurring. This same enthusiasm does not extend to women, who do not usually discuss their masturbation. So we're talking about a different acceptance of the same kinds of behaviors associated with sexual initiation and pleasure, which are distorted by what is learned about sexuality and are not openly discussed in the family except in the case of male adolescents.

These phenomena point to the need for dialogue and reflection. Clearly war, the economy and other issues are topics of discussion among men in the media and at the individual level. But why is it so hard to open a dialogue about something as commonplace as sexuality? Morality and the precepts it implies do not allow us to deal honestly with ourselves, or our pleasures.

Homosexuality in Cuba: Queer as Folk

Among groups of men who have been victims of any form of discrimination, homosexuals are at the forefront. In the 18th century, the first Cuban newspapers were already stigmatizing this sexual orientation. Since the 19th century, the term homosexual has unjustly condemned those who prefer this sexual variant to heterosexuality, which enjoys full acceptance; many heterosexual men boast of their hegemonic masculinity. Homosexuals are judged as lazy, weak, feminine and effeminately mannered. Such attributes would convey a lack of reliability for certain kinds of work, especially positions involving a powerful decision-making role. This behavior is generalized throughout Latin America and is deeply rooted in homophobia.

An issue as controversial in Cuba as homosexuality opens up a Pandora's box whose lid has been jealously guarded throughout Cuban history, as if it simply did not exist. *La maldición* (The Curse), written by researcher Victor Fowler, drew us into the topic from a historical-literary perspective. The author refers to a series of texts he considers to be foundational because they are from a period of genesis in our culture (Fowler 1998). Among these texts are those written by the priest José Agustín Caballero for the Havana newspaper *Papel Periódico de la Havana*, including his "Letter criticizing the male-female," on April 10, 1791, which identifies the issue of masculinity with male homosexuality:

Who can contain their laughter when they see a bearded man spend most of the morning combing his hair, getting dressed and admiring himself in the mirror like the vainest of women? (...) Truly, I don't know how any woman could accept this behavior from such cretins. These men play the role of roosters among women, by whom they deserve to be pitied, and of hens among men, by whom they should be despised. (Vitier et al. 1990: 75)

The Cuban philosopher Caballero designs a masculinity in which he associates the feminization of men with national defense:

I ask you now: If we needed to defend our country, what can we expect from such citizens or little narcissists? Do we know whether they will have the courage to face the hardships of war? Seneca said, "How can men who display a womanly and timid spirit be strong and forceful?" Let's not fool ourselves, men who are raised with music, dance, gifts, and such delights will inevitably degenerate into female behavior. (Vitier et al. 1990: 75)

If we keep in mind that these words were issued by one of our foremost thinkers, we see how a nation's masculinity is being constructed based on the exclusion of those who fail to meet these requirements. In the final verse of his text, he warns us about the danger to men of assuming feminine traits:

Wretched effeminate
you deserve this name
because you have
degraded your maleness.
Continue on the wrong path,
and consider a delight
your blatant foolishness
But you need not take offense,
If someone says to you in passing
"Best wishes, Lady Dionisia."
    (Vitier et al. 1990: 77-78)

If femininity in men implies rejection, masculinity in women is not without lesbophobic consequences. According to Italian philologist Analisa Mirizio, male clothing forms part of the sex role and, along with other factors, is a product of social learning (Mirizio 2000); it is normal for a man to wear men's clothing, but if a woman does, it is an affront to masculine virility and established morality.

It is likely that this prevailing opinion was the basis of a lawsuit filed on February 17, 1822 entitled, "Criminal case against Doña Enriqueta Favez for considering herself to be male and, dressed as such, deceiving Doña Juana de León to whom she was legally married," currently filed in the depths of the Cuban National Archives (ANC). Beyond the case itself, which was one of the most scandalous lawsuits ever seen in Cuba in the first half of the 19th century, it begs the question: what had Enriqueta[4] violated that caused her to be prosecuted? First and foremost: the public sphere of male power (González Pagés 2009).

Enriqueta Favez was a Swiss physician who, living in the village of Baracoa to practice medicine, dared to establish a lesbian relationship with a local woman named Juana de León. A review of the criminal case file reveals several contradictions in their unusual relationship, but what interests us is the analysis of masculinity and especially how the biological characteristics that defined her non-masculinity were being judged. The allegedly deceived wife stated, "'These anomalies forced me to spy on her until once when she thought I was asleep she was careless and I could see she had female breasts; they were not large but you could tell they had been used to breastfeed babies'" (González Pagés 2009: 55).

The fact that a woman of that era was a physician constituted a crime in and of itself. But that she should also violate the intentions of the Church and maintain a relationship that was considered unnatural made Enriqueta's trial a faithful replica of a Holy Inquisition tribunal by calling her a monster and a wretched creature, and unleashing upon her all manner of insults. In reality, more than judging the situation of victim-victimizer, the following framework was constructed to demonstrate the false masculinity of Enriqueta Favez, for which Juana declared:

> I hereby request that summary information from witnesses be admitted on my behalf and that those who present a statement testify to the best of their ability about the gender and physical impotence of the person who goes by the name of Enrique Favez. I also further request that the creature be brought to this city and in the presence of the court be examined by physicians and made to disrobe so that I may make deductions based on what I see, and that she remain jailed until otherwise determined and according to justice, proceeding without malice and as necessary. (González Pagés 2009: 57)

This is not a unique case in the Spanish colonial world. Other countries, including Colombia, also held trials for this reason, such as the 1745 case against two women in Popayán who were accused of "female sodomy." In Havana, 93 years after the Enriqueta Favez case, Puerto Rican writer Luisa Capetillo was arrested for wearing "clothing reserved solely for men" (Callejas 1915).

According to Professor Rodrigo Andrés, poststructuralist historians have established the fact that at different times in history there have been different valuations of gays and lesbians. The different discursive practices that not only name but actually create them have been very important for these views. Medicine and its "sanitizing" function in society was the cause of many of the controversies surrounding sexual diversity (Andrés 2000).

For example, in 1875 in Germany, a doctor by the name of Marx was one of the first scientists to request that sexual orientation be removed from the Criminal Code. He created a new term for this purpose, *Urnings*, which refers to a very special kind of person of male-female gender, thereby attempting to provide a medical justification for the phenomenon. For this scientist, the third gender seeks:

> Since childhood … society and girls' games; adults are distinguished by a female tone of voice and extreme timidity. Any little thing embarrasses or frightens them and causes them to blush; they abhor all strenuous exercise and instead relish needlework and show a strong preference for girls' activities, rings, necklaces, flowers and perfumes. They also display a persistent repugnance of women and would never want to have sexual contact with them. (Montané 1890: 581- 582)

This work was the subject of the most severe criticism in Cuba by Dr. Luis Montané,[5] who called it "moral depravity." During the First Cuban Regional Medical Conference, held in January 1890, his criticism of this research was particularly harsh:

> Is this the work of madman? Is it not, indeed, the opinion of Mr. Marx, who is considered a sage, a humanitarian philosopher? But in the end whether he is wise or mad is of little importance; what we need to remember is that his brochure has been freely sold in Germany and this demonstrates the existence of this shameful vice in that country. (Montané 1890: 582)

Later in his speech, Dr. Montané offered details about a study of homosexuality in Cuba involving 21 subjects, 4 of whom were European and 17 Cuban. The subjects were divided into active and passive groups according to their roles during sex, in order to emphasize the femininity of the latter. He went so far as to say, "Male prostitution is organized the same way as female prostitution. Furthermore, here are the names by which some of our pederasts are known: *Princess of Asturias*; *Passion*; *Verónica*; *Island Girl*; *Reglana*; *Camagüeyana*; *Manuelita*; *Albertina*; etc." (Montané 1890: 586).

The medical descriptions of some cases support this view:

> *Camagüeyana* has very unique ... buttocks, which are pushed together to form a single round mass ... in the case of M. Ll. (*Manuelita*) we observed ... the mucosal prolapse, forming two small regular lips joined at the bottom and classically reminiscent of a female dog's vulva ... *Camagüeyana* had covered his anus with a cotton cloth, probably to capture fecal matter – in their 'pursuit of femininity' some use this system to simulate a menstrual period. When we attempted to lift a corner of the cloth, the individual let out a piercing scream and fainted, and so we witnessed a hysterical Gran Mal seizure. Hysterical outbursts are commonplace in the world of pederasts! (Montané 1890: 588)

We know that one of the most discriminatory medical theories on 19th century women was the hysterical uterus theory, which essentially argued that women became hysterical because they had no penis. This same behavior was attributed to gay men as a form of discrimination that distances them from the possibility of being masculine and male.

Homophobia is alive and well in Cuban society and, like *machismo*, is rooted in cultural patterns. The efforts to socialize the debate around these issues found fertile ground with the premiere of one of Cuba's most famous films. *Fresa and Chocolate* (Strawberry and Chocolate, 1993), directed by Tomás Gutiérrez Alea and Juan Carlos Tabío, portrayed a gay male character as a courageous nationalist. This created a nationwide controversy in which the film met with the approval of the Cuban public.

Since then, other audio-visual projects shown on Cuban television have portrayed gay characters, sometimes in tangential roles and other times more directly. The same thing has happened in the theater, visual arts and other art forms. In fact, although reruns of Showtime Channel's English language series *Queer as Folk* have not been shown on Cuban television, many viewers have enjoyed these stories. Video rental operators have noticed customer interest in this issue and the series has been in high demand. *Tan raros como los demás*, the Spanish title of the series, is a call to rethink models of masculinity.

In some ways it is clear, at least in the artistic expressions of the 90s, that there is more freedom to express sexual diversity. This new trend seems to have left behind the contradictions of the 60s, 70s and 80s, when strong homophobic attitudes forced many homosexuals to leave the country.

The decade of the 60s will be remembered for events, which have yet to be studied in depth, that suggest the replacement of the traditional definition of masculinity inherited from the Republican and colonial periods with the redesigned "New Man." This new male model exposed the tension between masculinity and homosexuality that existed before and after 1959. In his study entitled *Los excluidos de la masculinidad: Espacio nacional y regulaciones sexuales en Cuba* (Excluded from Masculinity: National Space and Sexual Regulations in Cuba), Professor Santiago Esteso Martínez (1998) from the Universidad Complutense de Madrid, states, "This tension exposes - and we're talking primarily about masked form - a specific cultural ailment: the incompatibility of homosexuality with the law and the contract that governs social relationships and the construct of masculinity" (80).

In the second half of the same decade, Military Units to Aid Production (UMAP) were an example of the attempt to instill an apparent model of new or socialist masculinity. According to Carlos Monsiváis, the New Man was a "militant for the revival of Latin America" (in Esteso Martínez 1998: 87) who resisted all tendencies or ideas that might weaken him, including homosexuality. Some men taken to UMAP camps in 1967 spoke of the physical labor and the challenges they faced during their supposed transformation into New Men: "In the afternoon they want you to keep the same pace, but you just don't have the strength. The heat and humidity are good for the crops but draining for men" (Suárez Rivas et al. 2004: 5).

Thirty years after these events, well into the 21st century, the challenges of masculinity and its relationship to homosexuality have changed. There are no longer laws supporting homophobia, and activities promoted by government institutions have made the public visibility of homosexuals much more commonplace.

In his paper entitled *¿Masculinidad o masculinidades?: estudio con un grupo de hombres en una fiesta gay de Ciudad de La Habana* (Masculinity or Masculinities?: A study of a group of men at a gay party in Havana), sociologist Andrey Hernández (2007) emphasizes how the collective imaginaries among gay and straight men are constantly clashing:

> There is probably nothing as commonplace as discussing homosexuality. Whenever a group of straight men get together to hang out on the corner, regardless of their age, occupation, race or marital status, they inevitably start talking about "queers." One of the masculine ideals is to be successful, something that is impossible unless a man is morally irreproachable, which is only possible if he is straight. (39)

Another issue related to sexuality and masculinity is prostitution. According to sociologist Mairim Valdés, prostitution in Cuba has increased over the past decade, as have sexually transmitted infections.

In his thesis study, *Masculinidad y sexualidad: la bisexualidad y el VIH/SIDA en un grupo de hombres del municipio Plaza de La Revolución en la Ciudad de La Habana* (Masculinity and sexuality: Bisexuality and HIV/AIDS in a group of men from the Plaza de La Revolución municipality of Havana), when asked why bisexual men and HIV/AIDS should be studied, Valdés (2007) explained that this group was considered to be at highest risk among men who have sex with men (MSM):

> In the past twenty years, 76% of the men infected with HIV/AIDS nationally, and 90.5% of those in Havana, have been infected through what specialists call homo/bisexual sex, a category characterized primarily by the diversity and richness of the sexualities it includes. (67)

El jinetero (jockey) or *pinguero* (derived from the slang word for penis, *pinga*) are the terms used for a man who practices male prostitution in Cuba, a very controversial figure in social and family circles. In her essay entitled *Jineteros en La Habana* (Male Prostitutes in Havana), Rosa María Elizalde (1996) suggests:

> Except for the police, male prostitutes do not feel rejection in their community nor in the places where they practice their trade. This is in paradoxical contrast to the ethical precepts that support government and political institutions, and to the strong social rejection they were subjected to during the initial years of revolutionary government. (39)

The global participation of men in the sex trade, as workers, clients, or members of networks or mafias, has become a problem for many Latin American countries. The International Labor Organization has developed many projects, most significantly *Commercial Sexual Exploitation and Masculinity: A qualitative regional study of men from the general population*, a study containing information about the perception and awareness that men from Guatemala, El Salvador, Honduras, Nicaragua, Costa Rica, Panama, and the Dominican Republic have on the commercial and sexual exploitation of minors. This research is an important resource for implementing strategies to address this societal scourge, and also calls attention to hegemonic masculinity and its impact on male sexuality (Salas and Campos 2004).

References

Andrés, R. (2000) 'La homosexualidad masculina, el espacio cultural entre masculinidad y feminidad, y preguntas ante una crisis', in M. Segarra and Á. Carabí (eds.) *Nuevas Masculinidades*, Barcelona: Icaria.

Callejas, B. (1915) *Venus con pantalones*, La Habana: La Prensa.

Elizalde, R.M. (1996) *Jineteros en La Habana*, La Habana: Editorial Pablo de la Torriente Brau.

Esteso Martínez, S. (1998) 'Los excluidos de masculinidad: espacio nacional y regulaciones sexuales en Cuba', in *Dossiers Feministes 6. Masculinitats: Mites, De/construccions i Mascarades*, Valencia: Universidad Jaume I.

Fowler, V. (1998) *La maldición*, La Habana: Ciencias Sociales.

Ginart, B. (2003, June 3) 'Las marionetas más impúdicas', *El País*, Barcelona.

González Pagés, J.C. (2009) *Por andar vestida de hombre*, Bogotá: Red Iberoamericana de Masculinidades-Editorial Karisma. (CD-ROM)

Gutmann, M.C. (2000) *Ser hombre de verdad in la ciudad de México. Ni macho ni mandilón*, Ciudad de México: El Colegio de México.

Hernández, A. (2007) '¿Masculinidad o masculinidades?: estudio con un grupo de hombres in una fiesta gay de Ciudad de La Habana', thesis, Universidad de La Habana.

Mirizio, A. (2000) 'Del Carnaval al drag: la extraña relación entre masculinidad and travestismo', in: M. Segarra and Á. Carabí (eds.) *Nuevas Masculinidades*, Barcelona: Icaria.

Montané, L. (1890) 'La pederastia in Cuba', in *Primer Congreso Médico Regional de la Isla de Cuba en enero de 1890*, La Habana: Imprenta de A. Álvarez and compañía.

Morrisroe, P. (1996) *Robert Mapplethorpe. Una biografía*, Barcelona: Circe Ediciones.

Ravelo, A. (2005) *Intimidades, adolescencia y sexualidad*, La Habana: Editorial Científico Técnica.

Ruiz, D. (2001) 'Las rarezas del pene', *Men's Health, La Revista de los hombres*, April: 90-93.

Sade, M. (1990) *La Filosofía en el Tocador*, Barcelona: Tusquets Editores S.A.

Salas, J.M. and Campos, A. (2006) *Explotación sexual comercial y masculinidad. Un estudio regional cualitativo con hombres de la población general*, San José: Oficina Internacional del Trabajo. Online. Available HTTP: <http://webs.uvigo. es/pmayobre/textos/varios/explotacion.pdf> (accessed 16 February 2013).

Seidler, V. (2000) *La sinrazón masculina*, México: Editorial Paidós.

Suárez Rivas, R., Delgado, D., García, A., and Anh, L. (2004) 'Los hijos de Saturno', in *Graffiti*, La Habana: Facultad de Comunicación.

Valdés, M. (2007) 'Masculinidad y sexualidad: la bisexualidad y el VIH/SIDA en un grupo de hombres del municipio Plaza de La Revolución en la Ciudad de La Habana', thesis, Universidad de La Habana.

Verdú, V. (2003, June 16) 'El pene y su sombra', *El País*, Barcelona.

Vitier, C., García Marruz, F. and Friol, R. (1990) *La literatura en el Papel Periódico de la Havana 1790-1795*, La Habana: Editorial Letras Cubanas.

Endnotes

1   This is why the following terms, which connote high levels of stamina and strength, are used so often in Cuba to describe the penis: *palo* (pole), *tranca* (steel rod), *tubo* (pipe), cabilla, *macana* (baton), *mandarria* (sledge hammer), *cohete* (rocket), etc.

2   Nude dancers and theater performers in Cuba have always created controversy about whether they are necessary or just used as a way to attract larger audiences. In the case of the play *La Celestina*, it was very well received by theater critics and had one of the highest ticket sales in the past fifteen years.

3   Two of the many expressions used for male masturbation in Cuba. Matthew C. Gutmann, in his reports on the onset of masturbation in Mexican adolescents, found they use of terms like "putting a jacket on it" and "pulling on the rooster's head." Like Cubans, Mexicans believe that teen males should masturbate two or three times a day "because they have a stored semen" (Gutmann 2000).

4   The references in this chapter to Enriqueta Favez belong to a study which

was the source of the book *Por andar vestida de hombre* (For Dressing as a Man) (González Pagés 2009).

5   Professor Luis Montané was a prestigious specialist on 19th century Cuba. The Museum of Anthropology at the University of Havana is currently named after him.

Index